THE SCIENCE
OF THE
MANDALORIAN

THE SCIENCE OF THE
MANDALORIAN

THE ANATOMY OF A SPACE WESTERN

MARK BRAKE

COAUTHOR OF *THE SCIENCE OF STAR WARS*

Skyhorse Publishing

Skyhorse Publishing books may be purchased in bulk at special discounts for sales promotion, corporate gifts, fund-raising, or educational purposes. Special editions can also be created to specifications. For details, contact the Special Sales Department, Skyhorse Publishing, 307 West 36th Street, 11th Floor, New York, NY 10018 or info@skyhorsepublishing.com.

Skyhorse® and Skyhorse Publishing® are registered trademarks of Skyhorse Publishing, Inc.®, a Delaware corporation.

Visit our website at www.skyhorsepublishing.com.

10 9 8 7 6 5 4 3 2 1

Library of Congress Cataloging-in-Publication Data is available on file.

Cover design by Brian Peterson
Cover photo by Getty Images

Print ISBN: 978–1-5107-7059-1
Ebook ISBN: 978–1-5107–7060-7

Printed in the United States of America

This book is dedicated to Scrum.

CONTENTS

INTRODUCTION

"Study the science of art. Study the art of science. Develop your senses—especially learn how to see. Realize that everything connects to everything else."

—*Leonardo Da Vinci: An Exhibition of His Scientific Achievements* (1951)

Streaming Wars

We live in a Golden Age of television. It's not *just* down to the tech. Sure, it's true that there's been a revolution in media distribution technology. It's also true that there's been a revolution in digital TV tech, including TV streaming, HDTV, online video platforms, video on demand, and web TV. As a result, there's been a huge increase in the number of hours of available television. (During my childhood in 1960s Britain, we had but three channels. Television started at 5:15 pm, and the three stations would shut down before midnight!) More importantly, the current revolutions in tech have inspired a third revolution in content creation: *Stranger Things. Breaking Bad. Game of Thrones. Black Mirror. Sherlock. Watchmen. The Handmaid's Tale. Homeland. Westworld. The Queen's Gambit.* And many, many more.

The budgets of many of these television productions dwarf even the budgets of blockbuster movies. This trend met a peak in late 2022 when Amazon Prime announced that, having bought the television rights to *The Lord of the Rings* for $250 million in November 2017, Amazon was now committing to a five-season production worth at least $1 billion, making it the most expensive television series ever made.

This television revolution inspired Disney+ to enter the fray. Joining the likes of Amazon, Apple TV+, and Netflix, Disney+ attracted ten million user subscriptions by the end of its first day of operation. Disney had already acquired LucasFilm on October 30, 2012, for $4.5 billion. *Star Wars*, of course, was already a worldwide pop-culture phenomenon. The all-encompassing fictional Universe, created by George Lucas, was the fifth highest-grossing media franchise of all time (behind, believe it or not, Pokémon, Hello Kitty, Mickey Mouse, and Winnie the Pooh; it's a weird world in which we live.)

The combined box office revenue of *Star Wars* equates to over $10 billion, making it the second highest-grossing film franchise of all time (yet still *way* behind the Marvel Cinematic Universe). The list of highest-grossing films still has in its top fifty: *Star Wars: The Force Awakens* (at rank 4), *Star Wars: The Last Jedi* (at rank 16), *Star Wars: The Rise of Skywalker* (at rank 34), *Rogue One: A Star Wars Story* (at rank 38), and *Star Wars: Episode I—The Phantom Menace* (at rank 44).

Rather than being just a blaster from the past, *Star Wars* is more present than ever. The franchise's fantastically successful series of films have now been followed by multiple television series set in that same galaxy far, far away, and all airing across Disney+. The series span out from different anchors in the *Star Wars* timeline, and some conjure narratives that mix live action with animation.

The most popular *Star Wars* television series, and likely the firmest fan favorite for some time to come, is the franchise's first ever live-action series, *The Mandalorian*. Premiering as a launch title with Disney+, *The Mandalorian* traces the tale of the titular bounty hunter. Real name Din Djarin, this Mandalorian travels across the fictional galaxy with "The Child," real name Grogu, though affectionately known to *Mandalorian* fans as Baby Yoda.

Since its launch, *The Mandalorian* has been mostly regarded as one of *Star Wars*'s most engaging and exciting sagas. Disney has been commended for a most praised and highly acclaimed series, creating characters with gravitas and originality, worlds with depth and impact, resulting in some of the best *Star Wars* content ever.

Star Wars and Myth

Science fiction like *Star Wars* can be considered myth. Like the Greek myths of old, as well as the myths of King Arthur and the myths of the Old and New Testaments, these stories are not "actual" history, but nonetheless articulate some other kinds of truth. For example, there may or may not have been a historical Arthur. In a way, the question is irrelevant. What matters is that the myths surrounding his name represent something real about the Dark Ages in Britain. They're not lies. They represent a different kind of truth. They speak to us about important human values; about valor and kingship, about determination and loyalty, about doing the right thing, and about faith and renewal.

Science fiction can also be a form of modern myth. In his famous sci-fi novel, *Last and First Men*, British writer Olaf Stapledon tells the reader that his tales are an attempt "to see the human race in its cosmic setting, and to mold our hearts to entertain new values." Using science theories such as evolution and relativity, Stapledon suggests that attempts to imagine the human evolutionary future "must take into account whatever contemporary science has to say about [hu]man's own nature and his physical environment." As a result, Olaf Stapledon produced science fiction stories that took in the most recent concepts of cosmology and evolutionary biology. In this way, he created a new fusion of fact and fiction, a form of fable for a scientifically cultured twentieth century. In the words of Olaf Stapledon, the aim must not be just "to create aesthetically admirable fiction . . . but myth."

The Mandalorian is cast in a similar mold. The *Star Wars* galaxy presents a humanoid race in a cosmic setting. Science concepts, such as evolution and relativity and many more, are either strongly implied or indirectly referenced in the texts. What began as a modestly budgeted 1977 film made from a B-movie remix script has, astoundingly, burgeoned into myriad high-budget sequels, prequels, spin-offs, video games, and merchandise. *Star Wars* series like *The Mandalorian* permeate contemporary culture because they work so well as modern mythmaking.

Some commentators have criticized the sequel trilogy to *Star Wars*. The problem, they say, was that the sequel trilogy told, essentially, the same tale as the original trilogy: the Empire, demolished at the end of *Return of the Jedi*, is suddenly back on its throne and has to be vanquished all over again. Our republican warriors, conquerors at the end of *Return of the Jedi*, are popped back into their place as plucky rebels and outlaws.

Why? Because the myth that holds sway throughout *Star Wars* is the conflict between the individual and the overwhelming exterior forces of society, Empire, and galaxy. *Star Wars* also makes us compare our own highly technologized world with the *Star Wars* galaxy, which is a world *beyond* machines, a world where Jedi faith is as important as the certainties of science.

The status of *Star Wars* as a myth is very important. It's the reason the initial B-movie script went from Harrison Ford's "George, you can type this shit, but you can't say it" to globe-encircling success. It's all down to that mythic status. And the Janus-faced nature of the *Star Wars* myth is this: on the one hand, we have the aspect of individuality versus the dehumanizing authority of the Empire. On the other hand, we have the aspect of the place of the spiritual in an increasingly mechanized and materialist world. The success of *The Mandalorian* has both these myths embedded into it. The conflict between the individual and the overwhelming exterior forces of society is represented by Mando, an outsider and an outlaw. And the contrasting aspect of the spiritual in a mechanized galaxy is represented by Grogu.

Science and *The Mandalorian*

Is there an ideology to *The Mandalorian*? Or, more generally, since *The Mandalorian* exists within a specific fictional Universe, is there an ideology to *Star Wars*? By ideology here we mean an imaginary relationship to a real situation. In other words, how does the Mandalorian galaxy compare with our real Galaxy? The word "ideology" has something of a bad rep these days. People often take ideology to mean a misrepresentation of facts, but in this book, ideology is a necessary feature of understanding. That's because, should we lack an ideology, we would be badly disadvantaged in our perception of what happens in *The Mandalorian* and what happens in real situations. There is a real Universe, naturally, and it's way too big for us to know about in full, so we create an understanding by way of an act of the imagination. That, too, is an ideology. And it's a good thing. Call it worldview, call it philosophy, or call it science. Our ideology helps us filter the huge amount of information that daily pours into our collective consciousness. This torrent of intel, which flows into our minds, includes data from the sensory experience of watching or reading about *The Mandalorian* to mediated reviews of all kinds.

Here's how the ideology of science will help us with *The Mandalorian*. Science is a special kind of worldview. By way of its constant cross-checking with reality tests of different types, science helps us sharpen our focus. In our future together on planet Earth, some of what we see in *The Mandalorian,* and some of what we see in *Star Wars,* may well come to pass. But what might become true, and what not? Science is our best way of trying to work that out. But we won't forget that science, too, is subject to change. Science is not an ancient "truth," but a philosophy in flux; cumulative, constantly under repair, and always in use. And that makes science central to a most fascinating mission: to forge a future for us humans, and our planet, which is as bright and as bold as it can be for as many people as possible.

We won't be lame about the possibilities explored in *The Mandalorian*. After all, some of our current science theories may be shown in future to be false, while some related aspects of *The Mandalorian* turn out to be true. If you think I am being eternally optimistic, just consider something called the Falsification Principle. This principle, proposed by twentieth-century Austrian-British philosopher Karl Popper, is a way of telling proper science apart from pseudo-science. The Principle says that for a theory to be thought of as scientific, it must be able to be tested and conceivably proven false. For instance, the theory that "all members of Yoda's species are green" can be falsified by seeing a blue one.

The Mandalorian galaxy is furnished with all sorts of futuristic visions. A dazzling array of star systems and alien worlds, land-speeders and sub-light travel, beskar and bounty-hunters, droids and Krayt dragons. But *The Mandalorian* is part of *Star Wars*, and *Star Wars* is a science fiction franchise. Science fiction is simply a way of exploring the relationship between the human and the nonhuman. A way of saying something philosophical about the future of science and society in space.

The ideas of French philosopher Claude Lévi-Strauss may help here. Lévi-Strauss wrote about many things, but he focused particularly on the search for the underlying patterns of thought in all forms of human activity. And he suggested that myths, like *Star Wars,* are ways in which we work out the contradictions of our lives. These contradictions are clashes of opposites, such as the individual and society; the possible and the impossible; nature and culture; human and nonhuman. According to Lévi-Strauss, working through these contradictions is the very point of myths. And that's what we shall be working through in this book.

The Mandalorian and the "What If" Question

Comparing the ideology of science with the ideology of *The Mandalorian* will be helped by thinking about the so-called "what if" question. Science fiction has a famous faculty for imagining improbable things. But, then again, so too does science. Sometimes. Science fiction is a literary device for exploring imagined worlds, and in that sense is a kind of theoretical science. Scientists also make models of imagined worlds. They just happen to be more mathematical. They construct idealized Universes and set about tweaking the parameters to see what might happen.

The "what if" question is key to both science and *Star Wars.* What if we could travel freely through space? What if droids came to dominate the machine world? What if humans like the Jedi existed, and they could conjure near-magic through their use of the Force? Scientists also try to answer "what if" questions, but they are bound to stay within the confines of the known theories of science at that time. Fiction writers and movie directors have far more scope. And the very best science fiction can be used to ask profound philosophical or moral questions: the future of humans in space; the destiny of life in this Universe; the age-old question of good versus evil, and what those words even mean.

The job of this book is to compare the *Star Wars* future with our own potential futures, for there will be many, depending on what decisions we make about the future of our planet. To compare ideologies, Mandalorian and terrestrial, the book is divided into four sections: Space, Space Travel, Tech and Time, and Culture. Each of these sections is simply a way of exploring the future relationship between the human and the nonhuman, and the other contradictions of our lives.

I shall be using a very broad idea of what constitutes science. Communicating science has many practitioners, including scientists themselves, historians, moviemakers, sociologists, journalists, and philosophers. And the nature of science itself is characterized by many features, including science as an institution, science as a method, science as a body of knowledge, science as the key driver of the economy, and science as a worldview. I shall be using a very simple definition of science; namely, that science is a recipe for doing things. Science tells you how to carry out certain tasks, should you need to do them. This will help you understand why this book has a section on the science and culture of moviemaking.

Space

Space in *The Mandalorian* is a vast theater in which the planet-skipping stories unfold. But space is also a feature of the natural, nonhuman world, full of stars and alien societies. We compare Galaxy with "galaxy."

Space Travel

Having a huge theater of space is fine, but how do the stories' characters skip from one star system to another? This travel section includes questions of journeying to the stars, such as the huge interstellar distances, sub-light travel, and hyperspace.

Tech and Time

What might we humans one day become? Cyborg? Dark trooper? Genetically enhanced Übermensch, capable of incredible feats? This section looks at our evolutionary future.

Culture

A look at both the culture that triggered the creative imagination of *The Mandalorian* and how that created science fiction future culture might compare with our own.

Let's take a trip way beyond this Earth and into the infinite yonder, where your beskar armor will protect you from many things, but not the sight of a small, green, carnivorous humanoid with big black eyes and mysterious powers. This is the way.

PART I
SPACE

HOW DID *THE MANDALORIAN* REIMAGINE SPACE?

When I did *Star Wars,* I had an idea of doing this crazy 1930s serial action-adventure film, and the idea was it would be very, very fast paced and very exciting and the problem was there really were no special effects facilities at that time . . . So I was sort of forced to start my own company in order to make the movie. And that's really how I got started in the first place. I knew that I wanted something that was going to sort of, I had to push the limits of the ecology of the film medium in order to make this movie work . . . I needed to invent some new technology which was what we did at Industrial Light and Magic.

—George Lucas, interview, *American Film Institute* (2018)

The Sway of *Star Wars*

George Lucas's influence is everywhere in popular culture. For example, if we take another look at that list of the top fifty highest-grossing movies of all time, the rankings are run through with George's impact. At my time of writing, over half the list, twenty-six to be exact, are films within the genre of science fiction, all made since 1977:

Rank	Movie Title	Year
1	*Avatar*	2009
2	*Avengers: Endgame*	2019
4	*Star Wars: The Force Awakens*	2015

Rank	Movie Title	Year
5	*Avengers: Infinity War*	2018
6	*Spider-Man: No Way Home*	2021
7	*Jurassic World*	2015
9	*The Avengers*	2012
13	*Avengers: Age of Ultron*	2015
14	*Black Panther*	2018
16	*Star Wars: The Last Jedi*	2017
17	*Jurassic World: Fallen Kingdom*	2018
22	*Iron Man 3*	2013
24	*Captain America: Civil War*	2016
25	*Aquaman*	2018
27	*Spider-Man: Far From Home*	2019
28	*Captain Marvel*	2019
29	*Transformers: Dark of the Moon*	2011
31	*Transformers: Age of Extinction*	2014
32	*The Dark Knight Rises*	2012
33	*Joker*	2019
34	*Star Wars: The Rise of Skywalker*	2019
38	*Rogue One*	2016
42	*Jurassic Park*	1993
44	*Star Wars: Episode I—The Phantom Menace*	1999
49	*The Dark Knight*	2008
50	*Jurassic World Dominion*	2022

What's more, sitting at positions 15, 26, 47, and 48 are science fantasy movies which rely heavily on Lucas's late 1970s revolution in what we now call CGI: *Harry Potter and the Deathly Hallows: Part 2*, *The Lord of the Rings: The Return of the King*, *The Hobbit: An Unexpected Journey*, and *Harry Potter and the Sorcerer's Stone*, respectively.

In the summer of 2022, just after Marvel had released dates for a slew of new projects over the next few years, Disney+ dropped a bombshell. They reminded Marvel, and the rest of the world, just why modern cinema and television are able to conjure the magic they do. The six-part Disney+ documentary series, *Light & Magic*, chronicles the origins and evolution of Lucas's groundbreaking company Industrial Light and Magic (ILM). *Light & Magic* traces the tale of ILM, the visual effects company Lucas set up when making *Star Wars: A New Hope*. The series is a fascinating insight into how movie production was revolutionized. We watch interviews with current and former ILM employees and collaborators, and a story unfolds which pieces together a complete chronicle of the evolution of Lucas's empire from rather humble beginnings of miniature models and matte paintings, created back in the 1970s, through a movie-by-movie special effects roll call that details the blood, sweat, and tears poured into every little piece of the overarching project.

The end result is incredible. ILM has had an indelible impact on the art of filmmaking. As a LucasFilm company, ILM has served the digital needs of the whole entertainment industry for visual effects. ILM has been awarded three Emmy Awards, fifteen Academy Awards for Best Visual Effects, and thirty-three Academy Awards for Scientific and Technical Achievements. As well as the *Star Wars* movies, ILM has worked on films as diverse as (a small selection!) *Raiders of the Lost Ark, Back to the Future, The Last Temptation of Christ, Total Recall, The Godfather Part III, Jurassic Park, Schindler's List, Forrest Gump, Mission: Impossible, Titanic, Saving Private Ryan, The Bourne Identity, Harry Potter and the Chamber of Secrets, Pirates of the Caribbean: The Curse of the Black Pearl, Avengers: Age of Ultron,* and many more, as they say.

Imagining the Unimaginable

Light & Magic ends by bringing us up to date into the 2020s (in particular, with *The Mandalorian*'s revolutionary mix of practical sets and video screens). There's always been a major issue when writers and moviemakers have tried to imagine space. What weird and wonderful exoplanets exist in the depths of the cosmos? As *Star Wars* has evolved from 1977 to today, real science has discovered hosts of exoplanets, even in our local solar neighborhood. (Proxima Centauri b is the closest exoplanet to Earth, at a distance of about 4.2 light-years.)

So, *Star Wars* was right. There really *is* enough real estate in the sky, enough exoplanets going around other stars, to give every person on Earth, back to the first Stone Age human, their very own private, world-sized exoplanet. But moviemakers still have to answer questions such as, what will these exoplanets *feel* like to walk on? What's it like to *be* on such worlds? Lush and heavenly? Barren and hellish? Let alone the far more complicated question as to what cornucopia of creatures inhabit such exoplanets. How do moviemakers convey the taste, the feel, the human meaning of life on such worlds?

It's pretty tricky, imagining the unknown, and few creative minds in Hollywood and beyond have populated cinema and television with a more veritable zoo of exoplanets and aliens than in *Star Wars*. This series has been imagining alien worlds for almost fifty years. Desolate desert exoplanets like Jabiim and Jakku. Icy tundra worlds like Hothand, the exoplanet Mando leaves behind in "Chapter 1: The Mandalorian." Volcanic exoplanets like Nevarro. Forested backwater exoplanets like Sorgan. Green, but uninhabited, Earth-like exoplanets such as Tython. The gas planet of Endor, and its forest Moon of, erm, Endor. Yet the first confirmation of an *actual* exoplanet orbiting an ordinary star was made in 1995, almost twenty years after *A New Hope*.

The Mandalorian's Paradigm Shift

Given how hard it is to imagine the unimaginable when it comes to life on other worlds, *The Mandalorian* production team made sure they used the latest tech to render alien worlds as believable as possible. What form does this new technology take? In some senses, it's almost old school and traditional. It's a technique which has evolved for almost a century or so, in one form or another: displaying a live image behind the actors. *The Mandalorian*'s paradigm shift is in the way the production team used this technology. This revolution is predicted to help reinvent filmmaking and galvanize a new wave of content creators.

Meet the "volume," the largest and most refined virtual filmmaking framework ever made. Previously known as StageCraft, the volume is twenty feet tall, 270 degrees around, and seventy-five feet across. In film production, a volume is a space where motion capture and compositing happens. Volumes come in different, well, volumes! Smaller volumes are rather plain and simple affairs, where the motions of the actors behind computer-generated (CG) characters play out their roles. Other volumes are larger and more elaborate. They are built *into* sets; the kind you may have sometimes spotted in behind-the-scenes reels of *Star Wars* or *Marvel* movies. Normally, however, volumes hold one thing in common: they are bright green and blank workspaces. And they're *static*.

The Total Perspective Volume: Cleopatra and James Bond

Picture yourself in a volume. In a few moments, the director will ask you to convey through your acting, ultimately to a worldwide audience, just what the taste and feel of a scientifically alien environment is like, as you interact with a droid, a dark trooper, or a Krayt dragon (your job is admittedly made somewhat easier if you're playing a Mandalorian, as no one can see your face!). Your task is made all the harder by your immediate and actual working space. Green walls. Foam blocks that act as obstacles to be painted in later. And people wearing motion capture (mocap) suits, with daft-looking, pinned-on ping-pong balls adorning the suits and even their faces. Making matters worse is the fact that, in your working environment, everything that has caught a green reflection needs to be lit or colored out.

In the days of prequel-era *Star Wars*, advances enabled cameras to display rough previsualizations of what the final film would look like. CG backgrounds and characters were promptly substituted onto monitors. While these advances no doubt helped with composition and camera movement, it did less for the jobbing actor. The fictional world of the movie isn't suddenly and magically conjured up, as it might be with more practical sets and location shoots. It's useful to compare modern moviemaking with an old-school classic such as Joseph L. Mankiewicz's 1963 epic drama, *Cleopatra*. The film starred Richard Burton as Mark

Anthony and Elizabeth Taylor in the title role. The movie became the most expensive ever made up to that point, with estimated production costs of $44 million ($389 million in today's money). The costs nearly bankrupted the studio. Mankiewicz's location shoots were huge, and there were days when Twentieth Century-Fox had as many as ten thousand extras on set. Hollywood's craftsmen, talented people working long before the days of ILM, created monumental sets, particularly those for the scene in which Cleopatra arrives in Rome pulled by slaves riding a mammoth recreation of the Sphinx. In such circumstances, no matter the cost, you can see how it would be so much easier for the actor to feel, to some extent, that they were transported to the movie setting.

Pumping up the Volume

Unlike lesser volumes, *the* volume, built by ILM who still refer to it as StageCraft, is *not* static. Rather, the volume's backdrop is a set of gargantuan LED screens, the kind you might see onstage at rock concerts, but bigger. Bigger and, far more importantly, *smarter*. ILM reasoned that it's not enough to simply backdrop an image behind the actors. Moviemakers have been using projected backdrops since the silent era. Back in the day, backdrops looked pretty fake. Take the car chase in the 1962 James Bond movie *Dr. No*, for example. To fill in the backdrop, the Bond production team used rear projection (a moving image of the scenery projected onto a screen behind Bond in the car), but the end result looks rather lame. Sean Connery's steering doesn't seem to match the movement of his convertible. Previous volumes were fine (up to a point!) if the director just wanted a limited view out of a studio window or to fake a location behind a static shot. But modern moviemaking places far greater demands on production. Even something as simple, in fact, as merely moving the camera. Why? Because when the camera moves, it instantly becomes quite apparent to the audience that the backdrop is a rather flat and fake two-dimensional image.

The genius of StageCraft is twofold. For one, StageCraft, and other LED walls, generate an image which is shown live in photorealistic 3D by a powerful graphics processing unit (GPU). And two, the generated 3D scene is in direct symbiosis with the motion and settings of the camera, so if the camera moves to the left, the image alters in sympathy (just as it would in reality).

This is a very challenging trick to pull off. For starters, the camera needs to relay its real-time position and orientation to the GPU, which is, in essence, a monster of a video-gaming computer. The GPU then takes on board the data and renders it precisely in the 3D environment. Parameters such as perspective, depth of field, lighting, distortion, etc., are all tweaked accordingly. And, naturally, all of this must happen in the blink of an eye. The tweaked parameters must appear on the huge wall almost instantly. Should there be any kind of data delay, the motion of the backdrop would lag behind the camera by more than a handful of frames, and we'd be back to that James Bond car chase, and even the most inattentive viewer would notice.

The Mandalorian arrived at just the right time in production history. Just five years ago, it wouldn't have been possible to make *The Mandalorian*, to reimagine space in the same way. In the last half decade, a number of technological advances have been refined to get to this point: the rendering, the LED walls, the tracking, and so on. And all of these developments have meant the evolution of StageCraft, and the enabling of great variety. Be it alien exoplanet, spaceship hangar, New Republic prison ship, or temple interior, StageCraft can conjure it.

As with all such developments, innovation has many architects. Chief among them for our story is, of course, Jon Favreau. A long-standing driving force for this kind of filmmaking, Favreau bench-tested these movie methods on films like *The Lion King* before the ultimate tech tryout on *The Mandalorian*. Kudos go also to the progress in virtual filmmaking made by the likes of James Cameron and, in terms of work in motion capture, the sterling contribution made by industrious actors such as Andy Serkis.

The Volume's Bang for Buck

Such innovation doesn't come cheap. Estimates of the build cost of ILM's StageCraft range from $100 to $250 million. Hardly surprising, given that StageCraft is probably the most expensive and complex production environment ever constructed—but what a bang you get for your buck. What the likes of StageCraft add in tech overheads, they more than pay back in so many ways. Take on-location shooting, for instance. No longer necessary. And given that location shooting is wildly time-consuming, logistically mind-numbing, and prohibitively expensive, that's a big benefit. During preproduction, a virtual photography process was used to plan *The Mandalorian*'s filming, to work out what digital environments would be needed on set. These environments were then created by ILM and added to StageCraft, primed for live-action filming with the actors. Some of these environments were based on location photography in countries such as Chile and Iceland (no pun intended!), about which Favreau explained, "The actors aren't brought on location. The location is brought to the actors."

Fancy a trip to Tatooine? Instead of going to Tunisia to get those wide-open desert shots, you just fire up the volume. Of the $11 million it cost to make *Star Wars: A New Hope*, $700,000 of the total budget was spent on taking the film production to the North African desert, to give the movie that sense of real-world scope. Recall that situated in Southwest Tunisia lies the city of Tataouine, with the astounding architecture of the Ksars, the fortified villages of the Saharan inhabitants. George Lucas chose Tataouine as the base for his original film, as it sufficiently looked like a galaxy far, far away. But filming on location has more drawbacks than just the titanic budget costs. Lucas and his production crew accidentally heightened tensions between Tunisia and neighboring Libya by increasing activity in the area. One rather worrying report tells how matters between the two countries got truly tense when Libya demanded Tunisia immediately cease its deployment of a massive military vehicle near their shared border. The tension reached crisis point when Muammar Muhammad Abu Minyar al-Gaddafi warned that conflict was inescapable if Libya's demands were not met. Comically, the military ordnance in question was none other than the giant vehicle used as the Jawa Sandcrawler. To avoid war, Lucas had the massive prop moved to a more secretive location. The entire affair is exactly the kind of production headache rendered impossible by using the likes of StageCraft.

Instead of all that on-location expense, the volume's version of con-juring up a desolate desert exoplanet would be as follows. Rather than actually going to somewhere like Tunisia to capture those often-stunning desert landscapes with their unique coloration and native flora, you can simply build a sandy set and place a photorealistic desert behind the actors. That doesn't mean you can't send a small scouting production team to Tunisia. They can help create a visual experience unique to desert climates by collecting data. In the end, you have the best of both worlds, having captured a desert in high-definition 3D to be used as a virtual background (which would also make reshoots a whole lot easier!).

Imagine another example. Let's say you'd filmed a scene in Ireland. In fact, let's assume the scene was shot on the craggy island of Skellig Michael, situated off Ireland's west coast. (This is one of the most popular recent *Star Wars* movie locations, famous for scenes between Rey and Luke Skywalker in both *The Force Awakens* and *The Last Jedi*. This dramatic, twin-peaked and tiny island, with its ancient Gaelic monastery, was used as the Planet Ahch-To, where Rey goes to find Luke after his years of self-imposed isolation.) If you needed to change the dialogue and reshoot the scene over again, it would be far simpler. No need to travel back to Skellig Michael. You simply fire up your virtual background for the reshoots.

For *The Mandalorian*, digital environments were first designed by the show's visual art department. Then, during principal photography, the environments were rendered on a video wall in real time, allowing the filmmakers and actors to get a grip of their new virtual reality. ILM had previously used a smaller version of this tech for *Solo: A Star Wars Story* in 2018, but for *The Mandalorian* they used a twenty-one-foot-tall set, which was a full seventy-five feet in diameter, surrounded by a 360-degree semicircular LED video wall and ceiling.

The Manhattan Beach Studios set, referred to as the volume, is a tradi-tional name for motion capture stages. It had been Jon Favreau's original aim to merely use these stages as a way of providing realistic interactive lighting for actors, with a section of the screen behind them displaying a green screen so that a higher quality version of the background could be added in post-production. But during filming tests with this tech, the production team realized that the Unreal Engine (the world's most open and advanced real-time 3D creation tool for photoreal visuals and immersive experiences) could render visuals so quickly that they could have the background actually move with the camera in real time. This alacrity allowed the system to maintain the appearance of parallax (see more about this concept on page 72), where the environment appears differently, depending upon the angle it was being looked at, just as in 3D reality. While this effect causes some distortion to the image on the motion capture stages, it actually looks like a real environment when seen through a camera, so the images rendered on stage in real time were often of a high enough quality to be used as final effects when filmed on set.

Turn up the Volume

The volume has a host of backdrop and special effects that can be rendered ahead of time and shot in-camera rather than composited in later, which means a huge saving on time and money. The by-product of that, naturally, is a slicker and smoother creative process. In-the-moment decision-making by moviemakers and actors is made possible, as the volume is reactive to their requirements.

StageCraft is also light-years ahead on illumination. Lighting is now hugely simplified. The brilliant LED screen can deliver a dramatic wall of illumination, and given that it also represents the scene, the screen's lighting is precise and sympathetic to the needs of that scene. For example, a spaceship interior with a blue hue, or the iridescent smithereens of an electrical short-out, can show up just as blue on the actors' faces and the madly reflective helmet of Mando himself. Further refinement of light is possible, too. Placing a red light source high on the LED screen, but out of sight of the camera's "eyeline," can create an ominous portent of peril ahead on Mando's helmet.

Naturally, a volume also has some disadvantages. It has limits. Checking in at around twenty feet high, a volume is tall but not tall enough that a wide shot won't show the top of it, too. Viewers would see cameras and another kind of LED, as the ceiling, too, is a screen, albeit less powerful. In practice, this merely means liminal constraints on the camera lenses and angles that the team shoots with.

Another constraint is the size of the LEDs. In other words, the size of the pixels. This dictates how close the camera gets to the LED screen, and totally rules out zooming in on an object for a closeup. Too much zoom would result in moiré patterns, the patterns of stripes one sees on screens when a natural interference phenomenon appears on displays.

The Future of the Volume

The Mandalorian represents an historical high mark in the way moviemakers have used the latest tech to make alien worlds as believable as possible. While StageCraft is not the first application of LED screens (they've been available for some time on smaller scales), it's without a doubt the most high-profile demonstration of what's possible.

Nor is *The Mandalorian* a one-off. According to Miles Perkins, manager at film and TV for Epic Games, the makers of the Unreal Engine used in virtual production pipelines, as of late 2022 there are "roughly 300 stages, up from only three in 2019." These stage installations include ILM's StageCraft, situated at Manhattan Beach Studios, and used on *The Mandalorian*. Marvel also employed a bespoke StageCraft build in Sydney for *Thor: Love and Thunder*. Marvel was also heavily dependent on ILM's system at Pinewood Studios in London for *Ant-Man and the Wasp: Quantumania*. Pixomondo's Toronto-based LED stage has a long lease from CBS and has been used for seasons four and five of *Star Trek: Discovery*. Meanwhile, NEP's Lux Machina says that its virtual production credits include the Sony action movie *Bullet Train*.

Almost all production houses are now building, or evaluating, LED screens. The advantages are clear. Television productions can save money and look just as good. Movies can be shot on more adaptable schedules. And commercials won't be far behind; they too will find a way to use these walls. Not long from now, it will be uncommon to find a production that *doesn't* use an LED screen in some way. It will become the new norm.

The volume, in short, is what the future of entertainment may look like. Virtual production will transform film and TV content. Media will be produced in such a way that makes it more amenable to the metaverse or immersive experiences. Audiences will be invited to sit inside what they are experiencing. As long as moviemakers can construct a set surrounded by an LED wall, the resulting content can be given to an audience at home which they experience through a virtual reality (VR) headset. As Miles Perkins of Epic Games puts it:

> Once a team has iterated and finalized an asset, they can use it across mediums—linear content, experiential content, games, live events, and beyond. With a real-time game engine like Unreal, everything is so easily transportable that there's no longer a distinction between the needs of a linear deliverable versus an experiential deliverable. This means that virtual production is inherently preparing us for a new era of entertainment.

WHAT SIZE FORCE DID GROGU USE TO MAKE THE MUDHORN LEVITATE?

Physicists have traced three of the four forces of nature . . . But the fourth fundamental force, gravity, is different. Our current framework for understanding gravity, devised a century ago by Albert Einstein, tells us that apples fall from trees and planets orbit stars because they move along curves in the space-time continuum. These curves are gravity. According to Einstein, gravity is a feature of the space-time medium; the other forces of nature play out on that stage.

—Natalie Wolchover, *Why Gravity Is Not Like the Other Forces* (2020)

The Labors of Heracles

The Labors of Heracles are a dozen tasks that the greatest of the Greek mythical heroes, Heracles (whose name was later Romanized as Hercules), was told to carry out by Eurystheus, King of Tiryns. It all began when Hera, who hated Heracles because he was a living embodiment of her husband's infidelities, drove Heracles mad, leading him to murder his wife and children. Upon realizing the atrocity he committed, Heracles journeyed to the famous Oracle of Delphi to ask for penance. There, he was told to serve Eurystheus for a dozen years. If all his labors were completed, he would become immortal.

The dozen labors themselves are a beguiling bucket list of exotic jobs. They include slaying the Nemean lion, which dwelled in a deep, dark cave; the capture and return of Cerberus, the three-headed, dragon-tailed dog that was the guardian of the gates of the Underworld; and the capture of the Cretan Bull, a legendary creature that wreaked havoc on the island of Crete. Heracles did indeed complete the tasks and went on to join the superbly named Jason and the Argonauts in their attempt to obtain the Golden Fleece.

The Labors of Mando

In "Chapter 2: The Child of The Mandalorian," we have a labor of Heracles in miniature. The chapter tells the tale of the time Mando ran into some trouble, ended up in a cave, had an encounter with an exotic beast, and also came out unscathed from his labors. Mando is first ambushed by a trio of Vibro-ax wielding Trandoshan warriors. On besting the warriors and returning with Grogu to *Razor Crest* the following day, Mando discovers that a clan of Jawas are salvaging and looting his starship, as Jawas are wont to do. After a lengthy skirmish with the Jawas, Mando is ultimately stunned unconscious by their ion blasters. When he wakes, Mando is relieved to find the Child still in his cradle. But when they return to *Razor Crest*, which has been stripped of cables and parts, Mando finds that the ship has no power, and the two of them are stranded on planet Arvala-7.

As often happens in mythical Westerns, Mando and Grogu traipse across a desert. In time, they meet alien homesteader Kuiil, a member of the Ugnaught, a species of porcine sentient humanoids. Kuiil, a moisture farmer, is a little surprised to find Mando alive, but suggests that Mando trade with the Jawas. He has spoken. (Kuiil is essentially the kind of cranky old character one gets in Westerns, a mentor who helps our hero along on his journey.) They travel through the night and approach the Jawa clan the following morning. The trade-off starts badly, until an interpreter tells Kuiil and Mando that they will return the *Razor Crest* parts if Mando gets the enigmatic "egg" for them.

The Jawas grant Mando, Kuiil, and Grogu a ride with them in their sturdy sandcrawler, which comes to a halt near a rock formation. Mando and Grogu dismount and make their way through the rock formation until they reach a cave. Mando stands outside the cave and cautiously draws a weapon, as Grogu waits in his repulsorlift cradle. Mando bravely ventures into the cave. Taking a torchlight, at first he spies several ominously large bones, then quickly encounters a mudhorn. This is a gargantuan horned beast that looks much like Earth's woolly rhino, an extinct species of rhinoceros that was common throughout Europe and Asia until the end of the last glacial period.

The melee with the mudhorn doesn't start well. Mando is first driven out of its den. Once outside, Mando stands his ground and rushes to reload his blaster. But the mudhorn is like lightning. It charges forward and throws Mando against the mud. As Mando lies prostrate, the mudhorn eyes Grogu's floating cradle and lunges forward, but Mando makes sure the Child's cradle evades the attack.

The mudhorn switches its rage back to Mando, who decides it's flamethrower time. He directs a jet of flame at the creature and the beast is wounded, but not enough to take it out of battle. The mudhorn tries to retreat to the sanctuary of its cave, but Mando catches onto it with a grappling line. Once more enraged, the creature attacks Mando with its horn and hooves, and Mando is floored. Seeing the possibility of advantage, the mudhorn again charges at Mando, who unsheathes a blade, perhaps preparing for a final and fatal battle.

Grogu Intervenes with the Force

Ultimately, the Child comes to Mando's rescue. After intently watching the battle so far, along with the somersaults and beatings suffered by Mando at the horn of the beast, Grogu manages to muster up the energy to help Mando by using the Force. Mando looks exhausted, and rather pathetically draws a short and inadequate blade as the mudhorn makes to charge. Moments before the mudhorn gets to his prey, Grogu summons the Force, which appears to stop the beast in its tracks. Suddenly, having expected a fate that hadn't materialized, Mando looks up and is astonished (as far as we can tell from behind that helmet!) to find the mudhorn floating in midair.

As the creature treads air, Mando turns to see Grogu's intense focus as he keeps the mudhorn in stasis. Mando realizes that the Child is not a normal kind of creature, but wastes no more time and drives the short blade deep into the mudhorn's skull. Teamwork. A now victorious Mando approaches Grogu, who is overwhelmed by his efforts and falls asleep in the cradle. Mando goes back to the mudhorn's cave and picks up the egg from the mud and straw. Is Grogu justified in nodding off after his efforts? How much of an effort did he experience? In other words, what size force did Grogu use to make the mudhorn levitate, tread air, and render the beast vulnerable to Mando's attack?

The Labors of Newton

It's time to talk about gravity. For many people, the word gravity is synonymous with the name Isaac Newton. And for good reason. It was Newton who first created a mathematically quantified account of gravitation, one that embraced earthly and heavenly phenomena alike. In fact, Newton did such a good job in his 1687 book, *Philosophiæ Naturalis Principia Mathematica*, that his theory came to be seen, for more than two hundred years after as something close to the word of God. Even in the twenty-first century, when Newton's work has been subsumed as part of the broader canvas painted by Einstein's relativity theory, we still think in Newtonian terms. That's because Newton's laws still work well enough to guide starships to Arvala-7 and explain how a mere Child is able to levitate a mudhorn. (When US astronaut Bill Anders was asked by his son who was "driving" the Apollo 8 spacecraft carrying him to the Moon, he replied, "I think Isaac Newton is doing most of the driving now.")

One day (yes, his account of gravitation seems to have dawned on him all at once), Newton hit upon the grand theory that had eluded the likes of Kepler and Galileo. Newton's was a "system of the world," as he so grandly called it; a single, universal account of how the force of gravitation dictates motion in the whole of the cosmos. As Newton tells it in a memorandum from 1714:

> In those days I was in the prime of my age for invention and minded mathematics and philosophy more than at any time since . . . I began to think of gravity extending to the orb of the Moon and . . . from Kepler's rule of the periodical times of the planets being in

. . . proportion of their distances from the center of their orbs, I deduced that the forces which keep the planets in their orbs must [be] reciprocally as the squares of their distances from the centers about which they revolve: and thereby compared the force requisite to keep the Moon in her orb with the force of gravity at the surface of the Earth, and found them answer pretty nearly.

For those not well-versed in medieval English (let's be honest; who is?) Newton is saying that gravity is a field which is held in masses, such as that of the Earth, Moon, and planets themselves, and extends outward from them. His inverse-square law is a physical law that says the strength of gravity's field is inversely proportional to the square of the distance away from the source. In practice, this means that the farther away an object is from the source of the gravity, the less change can be observed in the object. So, gravity is a field by which a planet or other body draws objects toward its center. The gravity field also keeps all of the planets in orbit around the Sun, and so on. In short, gravity rules.

The Labors of Grogu

Consider gravity and the mudhorn. Why would the mudhorn land back on the ground if it ever managed to jump up high enough into the air? Why not just float off into space? The answer, of course, is gravity: an invisible field that means massive objects are drawn to one another, stay grounded, or fall. We can now think again about what Grogu has to do to make that mudhorn levitate. He "simply" has to muster enough of a counterforce to keep the creature afloat. He has to balance the weight of the mudhorn's bulk with an equal and opposite force.

First, we need to estimate the weight of that mudhorn. We can't keep on calling him "mudhorn," so let's name him Muddy. If Mando is good enough for a Mandalorian, then surely Muddy is good enough for a mudhorn. And, boy, that Muddy dude sure looks shredded. Given that muscle weighs more than fat, he probably comes in at a few tons. More precisely, recall that a little earlier, we mentioned the woolly rhino, an extinct prehistoric rhino species common throughout Europe and Asia.

Those dudes were also shredded. Like Muddy, they also weighed around 3.5 tons, or 7,700 pounds. That's some prime beef.

We need to convert that beef into Newtons, the scientific units for force (the Newton unit being, naturally, in recognition of Newton's work). The weight of an object is basically the effect of gravity on the object's mass. In the case of the mudhorn, the product of its mass and the local gravity field. Assuming the gravity field on Arvala-7 is the same as Earth's, we can calculate that Muddy's 7,700 pounds equates to roughly 34,250 Newtons (this seems reasonable, as the weight of a typical elephant on Earth is 19,600 Newtons, and Muddy is some beast).

Grogu needs to use the Force to conjure an oppositional force to Muddy's 34,250 Newtons. None the wiser? Perhaps it would help to think about what a single Newton would weigh and feel like. Let's use apples. Newton is said to have recalled, near the end of his life, that the idea of gravity came to him when he saw an apple fall from the tree in front of his mother's house. This story may be true. Newton's desk in his bedroom looked out on an orchard, and even Newton must sometimes have suspended his work to gaze out the window. The weight of a medium sized (100 gram/3 ounce) apple is one Newton. Thus, the counterforce Grogu needs to place under Muddy would be the equivalent of 34,250 medium-sized apples. Think that sounds surprisingly light? Think again. Or, even better, run along to the local mart and try lifting just a dozen apples.

If apples don't fire your rocket, how about burgers? It just so happens that one Newton is also equal to a little less than a quarter of a pound, such as the weight of a quarter-pound burger after it's cooked. It would seem insensitive to suggest that this burger unit be a mudhorn burger, but since the word "burger" has become associated with so many different types of sandwiches, ground meat in hamburger, buffalo in the buffalo burger, kangaroo, chicken, turkey, and even salmon burgers, there's no theoretical reason why we can't have a mudhorn burger. Let's hope Muddy can't read.

Anyhow, we can now recalibrate Grogu's labors, from apples to burgers, and say that the oppositional force Grogu needs to conjure from the Force is equivalent to 34,250 quarter-pound mudhorn burgers. If you showed restraint and wanted to be reasonably regular in your daily visits to the bathroom and ate only one quarter-pound mudhorn burger per day, it would take you almost ninety-four years to eat a total of 34,250 burgers. See what I mean? That's quite the Force Grogu is having to summon up in order to levitate that beast.

It's little wonder Grogu paid the price of exhaustion for coming to Mando's rescue. Yoda might be famous for saying "size matters not," but Grogu is a mere fifty years of age. Levitating the charging might of a mudhorn that weighs the equivalent of 34,250 quarter-pound burgers has simply got to take it out of you. That snooze in the repulsorlift cradle was well earned.

WHEN WAS *THE MANDALORIAN?*

This story happened a long time ago in a galaxy far, far away. It is already over. Nothing can be done to change it. It is a story of love and loss, brotherhood and betrayal, courage and sacrifice, and the death of dreams. It is a story of the blurred line between our best and our worst. It is the story of the end of an age. A strange thing about stories—though this all happened so long ago and so far away that words cannot describe the time or the distance, it is also happening right now. Right here. It is happening as you read these words.

—Matt Stover, *Revenge of the Sith* (2005)

The Mother of All Taglines

First, a quiz. This quiz is about movie taglines. Put yourself in the position of a tagline writer. Millions have already been spent on making a blockbuster of a movie. Hard work done by all. And now, it's your turn. The task, should you choose to accept it, and let's face it, you've already agreed to do this job, is to turn all that time and money into a simple punny phrase. In short, your tagline has to sum up the movie in a nutshell. Tagline writing is a delicate art unto itself. And the best are a balance of informative, compelling, and witty.

So, here's the quiz. How many of these ten superb taglines can you get right? Answers are upside down in the next paragraph.

1. "In space no one can hear you scream."
2. "Just when you thought it was safe to go back in the water."

3. "On every street in every city, there's a nobody who dreams of being a somebody."
4. "Be afraid. Be very afraid."
5. "Escape or die frying."
6. "The future is history."
7. "Whoever wins . . . we lose."
8. "On the air. Unaware."
9. "The last man on Earth is not alone."
10. "Check in. Unpack. Relax. Take a Shower."

Answers: 1. Alien; 2. Jaws 2; 3. Taxi Driver; 4. The Fly; 5. Chicken Run; 6. Twelve Monkeys; 7. Alien vs. Predator; 8. The Truman Show; 9. I Am Legend; 10. Psycho

Splendid examples, all, yet we know the mother of all taglines comes from that famous opening crawl which we first witness at the beginning of *Star Wars: A New Hope*. Against a Bible-black deep-space background, with just a mere sprinkling of stars, the crawl is preceded by the *Star Wars* logo, but at the very top of the title sequence is the opening static blue text: "A long time ago in a galaxy far, far away . . . ." This crawl is one of the most immediately identifiable factors of the franchise and has been frequently parodied. But it's that tagline which truly abides.

When Was *Star Wars*?

The *Star Wars* tagline begs a very interesting science question. When, in the history of spacetime, are the *Star Wars* stories, including *The Mandalorian*, meant to have happened? In March 1977, *The New York Times* printed a report that "scientists at the University of Chicago have determined that the Universe may be as much as twenty billion years old, considerably older than most current estimates." Today, the estimate of the age of the cosmos, assuming it all began with a Big Bang (and more about *that* later), is around 13.8 billion years.

Can we narrow it down any further? Can we perhaps define a time slot in the Universe's evolution when the *Star Wars* stories occurred? When we take a good look at the fabric of space in *The Mandalorian*, we can bear witness to the fact that, despite it being "a long time ago," the *Star Wars* galaxy shows evidence of maturity. The galaxy has millions of star systems, myriad inhabited worlds, and seemingly thousands of intelligent species capable of space travel.

The Cosmology of *The Mandalorian*

Looking at that famous tagline again, it's worth noting the wording a little more carefully. We are told that the *Star Wars* saga, of which *The Mandalorian* is part, takes place in a "galaxy far, far away." Not, in other words, another Universe entirely, nor in another spacetime, so we can assume that galaxy is in the same Universe as our Galaxy, within this realm of time. This single fact will help a lot in trying to pinpoint a period in the Universe's history when *The Mandalorian* might have happened.

Our standard cosmological model is the Big Bang theory. The Big Bang describes how the Universe began, how it evolved and is still evolving, and speculates as to how it might one day end. According to theory, Galaxies first formed about a billion years after the Big Bang itself. So, if the Big Bang theory is correct, the galaxy of *The Mandalorian* cannot have formed in the first billion years. Like *Star Wars*, *The Mandalorian* implies that there are thousands of star systems, some of which we witness, and many of which harbor mature planetary systems teeming with sophisticated and intelligent life. Our own Solar System formed and matured over a period of roughly five billion years, so it's safe to assume that the planetary systems we see in *The Mandalorian* formed five billion years after the formation of the galaxy.

The Mandalorian Clock

Keeping a time tally, so far *The Mandalorian* clock has ticked one billion years for the formation of the galaxy, and roughly five billion years for the formation of the planetary systems. A total ticking of six billion years to date.

Consider the cornucopia of multicellular creatures we meet in *The Mandalorian*: Blurrgs and Banthas, Devaronians and Dewbacks, and of course Krayt Dragons and Mudhorns, beasts of different shapes and types.

On Earth, it took around two billion years of evolution for single-celled organisms to develop into multicellular organisms, and another billion years or so until the beings began to take a form we would recognize as animals. That's roughly three billion years added to our *Mandalorian* clock.

But, who knows, the speed of life might be different in *The Mandalorian* and in the Star Wars Universe in general. The trouble is, biology is parochial. In the entire Universe, so far, we only have one biology to study, namely terrestrial biology. And, as we have nothing with which to compare terrestrial biology, it's tricky coming to any lasting and general conclusions. Consequently, we often use other sciences to try nailing down better hypotheses.

Imagine the creation of fictional Mandalore, to take just one planetary example. Given that we know life is meant to have evolved on Mandalore, we can assume it sat warmly and happily in its Goldilocks Zone, orbiting its parent star. Mandalore was a terrestrial planet once home to jungles, lush forests, grasslands, and lakes until war reduced it to a mere desert. We can assume Mandalore had all the common ingredients for life, or at least the ingredients we believe worked on Earth, according to our current science. Will the same ingredients work in other worlds? Does life's pathway always move at the same rate, or is it possible to speed up life and enjoy evolution in a faster lane?

It's an interesting question. Assuming life evolves elsewhere in the Universe, we know that life's pathway, evolution, depends on many things. From bacteria to Blurrgs, size is certainly a factor. Changes in the planetary environment, such as asteroid strikes and climate changes, will also influence the speed of life. We don't see too many species on *The Mandalorian*, as there's only so much one can do on a limited budget. But we know there are a lot more species roaming the Earth. Estimates vary from two million to one hundred million species. If we ignore bacteria, our best guess is about ten to fourteen million species.

Like on Earth, the time scale for evolutionary change on Mandalore would have been very long. On our planet, a typical period for the appearance of one advanced species from another is about one hundred thousand years. Biologists believe that humans branched off from their common ancestor, chimpanzees, about five to seven million years ago.

In that time, human subspecies like Neanderthals developed and died. As humans are meant to exist on Mandalore, they surely experienced a similar story.

Let's also remind ourselves that there may be speedier types of life out there in space. Consider bacteria. They are life's great success story on Earth. There are more bacteria in the gut of each person alive than all the humans that have ever lived on this planet. They occupy a wide variety of habitats. They have a broader range of chemistries than any other group. They are adaptable, indestructible, and astoundingly diverse. And their speed of life has been long and ancient.

Let's conclude this section on *The Mandalorian* clock. This is, naturally, only an approximation, one which assumes similar pathways in galactic, planetary, and biological evolution in the *Star Wars* galaxy as our own Galaxy. Nonetheless, our *Mandalorian* clock tally reads six billion years for a planet on which life can develop, and a further three billion years until evolved beings begin to take a form we would recognize as animals. Roughly nine billion years.

All told, this means that the drama that unfolds in *The Mandalorian* needs to be around nine billion years after the Big Bang. Since the Earth is roughly 4.5 billion years old, and the Universe approximately 13.8 billion years of age, this puts the age of the *Mandalorian* story *very* roughly around the same age as planet Earth. A truly ancient tale, and fully deserving of the tagline, "A long time ago in a galaxy far, far away . . . ."

Big Bang Spanner

The chief caveat on our clock conclusion, of course, is that mulling over where *The Mandalorian* sits in the history of our Universe is way trickier than trying to pinpoint where *Star Wars* planets might be in the history of their evolution. But there's a much more prodigious problem. What if our standard cosmological model is wrong? What if the Big Bang theory is broken?

Consider this recent cosmological controversy. In the summer of 2022, the James Webb Space Telescope (JWST) reported some interesting findings from the edge of the Universe. Launched on Christmas Day 2021, the JWST, a NASA project partnered with the European Space Agency and the Canadian Space Agency, is the largest optical telescope in space. It's mission includes observation of the "first" stars and the formation of the "first" Galaxies. And therein lies the problem. Those very first results from the JWST appear to show that massive, luminous Galaxies had already formed within the first 250 million years after the Big Bang. That's way earlier than the billion years it's meant to take for the first Galaxies to form, according to the Big Bang theory. This has repercussions for our *Mandalorian* clock.

Mere days after JWST's first science data became available, teams of astronomers presented their analysis. One team announced an unexpectedly massive galaxy corresponding to an age just 225 million years after the Big Bang, and another even claimed that some candidate galaxies were "only" 180 million years after the beginning of the cosmos. If confirmed, such findings would seriously challenge current cosmological thinking.

"The Big Bang Didn't Happen"

Some scholars went further. They suggested that, rather than JWST shedding light on the wonders of the Universe, the mission has instead called the very origins of the cosmos into question. On seeing those first images from the JWST, Allison Kirkpatrick, an astronomer at the University of Kansas, said "I find myself lying awake at three in the morning wondering if everything I've ever done is wrong." She was far from being the only voice of doubt. The JWST findings were also pounced upon by scholars with an axe to grind about the Big Bang theory. Chief among these was Eric Lerner, a popular science writer and independent plasma researcher. In August 2022, Lerner penned a piece called "The Big Bang didn't happen." The article went viral.

In this piece, Lerner goes for the Big Bang's jugular: "It has now become almost impossible to publish papers critical of the Big Bang in any astronomical journals. It is time to end the censorship and to let the debate begin. Cosmology can emerge from its crisis once it is recognized that the Big Bang never happened." Regarding the JWST findings, Lerner says that the Galaxies captured by the JWST are too ancient and too numerous to be compatible with the Big Bang theory. Why? Because, he says, it's impossible for Galaxies as large as our Milky Way Galaxy to form in a mere couple hundred million years or so. The evidence, Lerner concludes, points not to an expanding Universe, but a nonexpanding one. Basically, there was no Big Bang.

Cosmology in Crisis

Obviously, if Lerner's hypothesis turns out to be true, this would have a huge impact on the timing of our *Mandalorian* and *Star Wars* clocks. Our calculations would prove rudderless in time, as we would have no starting point from which to actually calculate *Mandalorian* galactic and planetary evolutions. In short, no beginning.

Cosmology has been in crisis before. In fact, a cosmological crisis led to the Scientific Revolution in the first place. Here's the backstory. In what follows, and we go into some considerable detail, keep in mind the contemporary cosmology of the Big Bang, as we read how wrong cosmological models were in the past.

In the world before the telescope, humans relied on naked eye observations to work out what might be happening in the sky above. All primitive but intelligent cultures would have done the same, whether on Earth, on Coruscant, or on Mandalore. For thousands of years, avid sky-watchers would have scanned the dark nights and mapped the heavens. Wherever they were, within reason, they would have noticed a number of things. The entire sky appears to rise in the east and set in the west, so it's natural to assume, from this optical illusion and knowing no better, that the Universe is revolving around your home planet.

Being at such huge distances, the stars themselves remained relatively stationary. They were thought to resemble pinholes in a dark fabric through which was glimpsed a cosmic fire beyond, so the idea of a solid sky evolved, the so-called firmament of the Bible, with the stars mere holes in the firmament through which the light of God's heaven shone. But the real challenge came with the planets. Not that they were actually *called* planets back then. The planets, again whether on Earth, on Coruscant, or on Mandalore, would have resembled vagabond stars, and they would have beguiled and bewitched sky-watchers for many generations. The planets were small in number. On Earth, only seven of these "wanderers" (the meaning of "planets") were to be seen among the thousands of star-lights that bejeweled the sky. On Coruscant and on Mandalore, maybe something similar. Without the aid of an optical device, the "wanderers" looked the same as the stars, but their behavior was quite different. True, like the stars, Mercury, Venus, Mars, Jupiter, and Saturn (as well as the Sun and Moon, which were both considered planets) all seemed to revolve once a day around the Earth. But the planets also had a peculiar motion. The reality of this motion, of course, derives from the fact that the entire local planetary system is actually in orbit around the Sun, and all the observations that the Earth-based, third-rock-from-the-Sun, ancient sky-watchers made were biased by this fact. (Again, we can imagine similar historical issues with sky-watching on Coruscant or on Mandalore.)

These seven "planets" wandered and looped along their path across the night sky, yet they didn't roam around the entire sky. Their weird behavior was limited to a narrow strip of sky, a belt that encircled the spinning globe of the Earth at an angle of about twenty-three degrees to the equator. This belt, the Zodiac, was separated by the ancients into a dozen parts, and each part named for the constellation of fixed stars in that region of the Zodiac. Along this belt, the planets roamed.

Cosmology: A Lesson from History

From these observations, an ancient Earth-centered cosmology was born. (We can easily imagine a Mandalore-centered cosmology, too.) The first theory of the structure of the Universe emerged. It's possible that ancient astronomers on Coruscant and Mandalore made the same mistakes. According to this Earth-centered system, the planets moved, with the same regularity as that of the rotating sphere of fixed stars beyond, in circular orbits around the central Earth. This gave a good account of observations of the Sun's behavior on its yearly journey through the plane of the ecliptic. Their system also gave a reasonably accurate account of the rather less regular motion of the Moon. But, and here's the chief scandal, the circular orbits got nowhere near explaining the observed motions of the other five wandering planets.

Nonetheless, once the position and shape of the orbits were established, it was possible to make an educated guess on the layout of this ancient Solar System. Planets such as Jupiter and Saturn described a slow motion across the sky, appearing to keep up with the fixed stars beyond. Today, we know that it takes Jupiter roughly twelve years and Saturn around thirty years to make a complete journey around the central Sun. To the ancient eye, since these planets very nearly kept pace with the stars, they were assumed to be far from Earth, and close to the stellar sphere, which bounds our Universe.

Figure 1. The geocentric or Earth-centered Solar System.

In contrast, the Moon loses twelve degrees each day in its apparent race with the stars. The ancients must have thought this justified the suggestion that the Moon was closer to the Earth, which was assumed to be stationary, as well as at the center of the system. So, the outer limit of this cosmos was the stellar sphere, and just inside was Saturn, since it was the planet that took the longest to move around the Zodiac. Next came Jupiter and Mars, arranged in order of decreasing orbital period, the time taken to make one complete orbit. Innermost was the Moon, since the lunar orbit placed it closest to Earth.

The remaining three planets, the Sun, Venus, and Mercury, posed a problem. All three vagabond stars made their apparent journey around the Earth in the same common time: one year. Their order could not be unpicked with the technique used for the other planets. Indeed, there was much disagreement on this matter among ancient philosophers. The layout shown in the diagram above—Sun outside, followed by Venus and Mercury within—was not undisputed, but for reasons now lost in the mists of time, was the most popular order.

A Cosmology That Worked

The layout of this ancient cosmology held such remarkable potency and power that it lasted more than 1,500 years, up to the days of Galileo. Consider the science communication of this cosmology. Figure 1 shows almost all the knowledge enjoyed by the nonastronomer in ancient times. The phrase "inferior planet" was used for those planets (Mercury and Venus) that were between the stationary Earth and the orbiting Sun, and the phrase "superior planet" was used for those planets (Mars, Jupiter, and Saturn) lying beyond the Sun's orbit. The system gives no indication of the dimensions of the orbits, no account of the irregularities of the planets in motion, but these further developments of ancient astronomy were too mathematical for most most people to grasp.

Consider astronomical research. This cosmology proved a potent tool, at once efficient and fertile. For one thing, the notions embodied in Figure 1 gave a satisfying account of both the phases of the Moon and lunar eclipses. The ancients had used these eclipses as part of their reason for believing that our world was spherical; the shadow cast by the Earth upon the Moon during these events was, after all, round, and the only solid that always projects a round shadow is a sphere.

Philosophers later used the concepts of this cosmology to make measurements of the Universe that were quite startling in their accuracy. For example, the Greek thinker Eratosthenes measured the Earth's circumference with unerring precision during the third century BC. During the second century BC, Aristarchus, another Greek thinker, brilliantly calculated the sizes and distances of the Sun and Moon. So, even though the cosmology was in error at a fundamental level, the system still had power and ingenuity.

In short, the Earth-centered cosmology was a simple but potent model of the Universe that worked, up to a point. Could the same be true about the Big Bang? The Earth-centered model made common sense to simple naked eye observations that could be made nightly, cloud cover allowing, as the sky seemed to revolve around the central Earth. This cosmology could also be used to predict important events such as eclipses, those ominous cosmic encounters which had previously given the ancient mind such pause for superstitious thought. Thus, it's hardly surprising that the model became a rational tool, a conceptual scheme with an increasing hold upon the minds of both astronomers and ordinary folk.

A Hitch in the Heavens

The Earth-centered cosmology had trouble explaining, to various degrees, the annual motions of the seven wandering "planets," the Moon, Mercury, Venus, the Sun, Mars, Jupiter, and Saturn. The Galaxy of rotating stars proved no problem. They seemed to never alter their fixed place relative to one another, or to the observer on Earth. As an assumed solid sphere, they rotated en masse, or at least appeared to do so, due to the Earth's rotation on its axis.

In contrast to the eternal dependability of the sphere of fixed stars were the tramp stars, the planets. Given the relative satisfaction, if not succor, to be gleaned from the regularity of the constellations, how unsettling the wandering stars must have been, especially to philosophers trying to test their wits against the skies. Thank heavens for the Zodiac. At least the tramp stars stuck to this same belt along the firmament.

It's not easy, building a true mental model of the sky. Allow yourself to be transported back to the ancient world on Earth, or, by analogy, Coruscant or Mandalore. Your mission is to prove, with the information available to your naked eye, that your "world" is indeed heliocentric (Sun-centered) rather than geocentric (Earth-centered). To the naked-eye observer, the entire canopy of the heavens, stars, planets, and Milky Way, appears to rotate around a central Earth, rising in the east and setting in the west. The geocentric model explains this day-to-day procession.

The major problem with the basic geocentric system is that it simply didn't work. The Sun and Moon are reasonably well behaved. They move smoothly, in a reasonably regular fashion, along the zone of the Zodiac. Would that the same were true of the other five planets. Alas, they weave their wicked way, ambling along the sky, west to east, the same bearing as the rest of the heavenly traffic. Then, at intervals, they slow down and stop, and begin winging their way in the opposite direction. As if that isn't bad enough, they then seem to have another change of heart, turn around, and resume soaring in the original direction. Their behavior is hardly consistent with motion in a circle, with the Earth as the center.

What's more, two of the planets, Mercury and Venus, seem to have some kind of unhealthy obsession with the Sun, forever sticking close, like paparazzi to a princess. True, they may often appear to race ahead, or hang behind, but their general mode is one of pestering. Little wonder that when Venus rose ahead of the Sun she was known as Phosphorus, the "morning star," and when she set behind the Sun was known as Hesperus, the "evening star." Imagine what a huge advance it was when Greek thinker Pythagoras, according to legend, unraveled the mystery; he was the first to realize they were the same planet.

Geocentrism: On Earth and on Mandalore?

Aristotle's vision of the cosmos held sway for almost two millennia. Aristotle's cosmology was of a two-tier, geocentric Universe. The Earth, mutable and corruptible, was placed at the center of a nested system of crystalline celestial spheres, from the sublunary to the sphere of the fixed stars. The sublunary sphere, essentially from the Moon to the Earth, was alone in being subject to the horrors of change, death, and decay.

Beyond the Moon, the supralunary (celestial) sphere was immutable and perfect. After all, the heavens did not appear to change to the sky-watcher. Crucially, the Earth was not just a physical center. It was also the center of motion, and everything in the cosmos moved with respect to this single center. Aristotle declared that if there was more than one world, more than just a single center, elements such as Earth and fire would have more than one natural place toward which to move. In his view, this was a rational and natural contradiction. Aristotle concluded that the Earth was unique, and his cosmology became the classic geocentric system of the ancients, with the most important proviso being this: the planets must be moving in regular motion in a perfect circle.

The Greatest

Aristotle's flawed cosmology was "perfected" even further by antiquity's last able astronomer, the Roman polymath Claudius Ptolemy. Ptolemy's magnum opus, *Almagest*, which means "the greatest," was the only surviving comprehensive ancient treatise on astronomy. Written in the second century AD, *Almagest* consisted of thirteen books, a state of the Universe report on Aristotle's cosmology, and a guide to the complex motions of the stars and planetary paths. Hardly a pocket guide! Within its pages, Ptolemy claimed to marshal a model based on the observations of his predecessors, spanning more than eight hundred years.

Ptolemy had a crack at solving the problem of the planets and their insistence on tormenting sky-watchers with their weird wandering motion. But his model was even more divorced from reality, a rather farcical fairground of a Universe, wheels within wheels, like one gigantic wind-up toy, complete with clockwork components. In short, Ptolemy used epicycles.

As a master mathematician, Ptolemy took up the ancient philosophical mission to excuse the evident irregularities of the paths of the planets across the ancient sky. In pursuit of heavenly perfection, Ptolemy, like the geocentrists before him, had the planets rotate in circular motion at uniform speed. But his "solution" to the problem of the planets came in the form of a rather ridiculous but cunning proposal: the notion that the planets were attached to circles attached to concentric spheres. Thus, the weird motion of the planets is explained away by sleight of mathematics.

That an ancient astronomer with the talent of Ptolemy could convince himself that this elaborate scheme still amounted to uniform circular motion is a testament to the influential curse of geocentrism. In particular, the three powerful ideas, completely wrong, but deeply ingrained into the mindset of ancient astronomers: that a stationary Earth sat at the squalid center of the Universe; that heavenly objects were made from perfect material, unable to change their intrinsic properties, such as brightness; and that all motion in the heavens was uniform and circular.

Ptolemy's new model even explained, to some extent, the more anomalous and occasionally spectacular celestial events in his Earth-centered system. Though the weird motion of the planets might still be somewhat problematic, lunar and solar eclipses might seem incredible, and episodic changes in a body's brightness might trouble, the canny gadget of Ptolemy's epicycle was simply the slight adjustment needed to explain away the glitches.

The Revolution

There's no getting away from the fact that, in the hands of the likes of Aristotle and Ptolemy, astronomy had become an abstract sky geometry, divorced from reality. Nonetheless, from antiquity on, hardly anyone doubted the geocentric model. If the abstractions did not precisely fit observation, reason and common experience condoned it. Medieval philosophers and poets, encyclopedists and educators, noblemen and kings, all spoke of the Universe much as Aristotle and Ptolemy had described it back in ancient times.

A revolutionary overhaul of the skies was well overdue, and the man who started that revolution was Polish astronomer Nicholas Copernicus. In time, his clear and detailed explanation of a rotating Earth in a Sun-centered system shook the world. Armed with the inspiration of newly edited texts from the ancients and a strong aesthetic sense, Copernicus dared to center the Universe around the Sun. He had the courage to confront common sense, a sufficient skill in science to make the Sun the star of his heliocentric system, and, as a Renaissance humanist, enough incentive to bring the whole edifice of ancient thought plummeting down.

Terrified of the social, political, and religious repercussions his theory would inflame, however, the Copernican model was not set out in its final form until its creator lay on his deathbed in 1543. For the first time in written history, the Sun and its planets were set out in the correct order, even though the main reasons for Copernicus's revolutionary change were philosophic and aesthetic.

AD 1543 marks one of the great turning points in human history. It marks the beginning of the Scientific Revolution, and the modern age. Copernicus's great book, *De Revolutionibus Orbium Coelestium*, placed the Sun at the center of our planetary system, and heretically downgraded the position of the Earth to that of mere planet. Copernicus set in train a revolution. A new physics was born, and a new mantra: if the Earth is a planet, then the planets may be Earths; if the Earth is not central, then neither is humanity.

Paradigm Shift

You may well be asking, how is this relatively ancient cosmological crisis relevant to today's Big Bang and our calculation on the *Mandalorian* clock? To answer that question, consider the works of American philosopher of science, Thomas Kuhn. Kuhn made his mark with his analysis of the progress of scientific knowledge. In two books, *The Copernican Revolution: Planetary Astronomy in the Development of Western Thought* (1957) and *The Structure of Scientific Revolutions* (1962), Kuhn developed the idea that scientific fields undergo periodic "paradigm shifts." As with geocentrism, when one worldview, or paradigm, gives way to another, all scientific challenges are met from within the boundaries of this new framework. And, importantly, the notion of scientific truth, at any time, cannot be established solely by objective criteria, but is defined by a consensus of a scientific community. In short, real scientific truth is what scientists *say* it is, whether geocentrism or Big Bang.

In Kuhnian terminology, ancient cosmology needed a paradigm shift. The paradigm of geocentrism flew in the face of the facts implied by Copernicus's new and alternate heliocentric paradigm, which for some considerable time was just out of reach. The paradigms were incommensurable; they were competing accounts of reality, which cannot be coherently reconciled. The same goes for Big Bang cosmology. The conservative nature of consensus means that Big Bang cosmology will have to be truly exhausted of its coherence before scientists give it up for something else. Remember, both geocentrism and the Big Bang had aspects to their cosmologies that worked, even when the anomalies begin to grow.

If you Google reactions to those first images from the JWST, at first, there were plenty of lay reports that the Big Bang might be in trouble. But pretty quickly, the big guns of the science establishment, the likes of Neil deGrasse Tyson and Brian Cox, were shoring up confidence in the old Big Bang cosmology once more. So much so that it's now pretty tricky finding an alternate point of view on the matter. And that's kinda how science works.

Finally, here's our answer on "when was *The Mandalorian*?" If Big Bang cosmology remains intact as a theory, then the *Star Wars* story is *very* roughly around the same age as planet Earth. But if the Big Bang is to eventually go the way of ancient geocentrism, then the stories could be as old as forever, lost in the mists of history, a very fitting conclusion for a tale that unfolded "a long time ago in a galaxy far, far away . . . ."

DID YOUR ANCESTORS MUNCH A MUDHORN?

The mudhorn was a large horned creature that inhabited Arvala-7. Mudhorns had a large horn, flat teeth, and long, woolly fur. It was oviparous, laying a single egg per clutch. The egg was huge and was cherished by Jawas. The exterior of this egg was woolly in the same manner as the skin of the mudhorn, and the interior yolk was yellow.

—*Star Wars* canon (2023)

The Life and Times of the Mudhorn

As we already know, we first meet the mudhorn in "Chapter 2: The Child of the Mandalorian" when Mando does battle with the beast. They appear to have been large and non-sentient creatures whose chief habitat was the desert. They also appear to have had gray skin, brown fur and eyes, a horn (evidently!), and spikes along the ridge of their backs. There appears to have been a smaller kind of mudhorn too, a kind of mudhorn lite, that just had the gray skin and a horn.

The mudhorn appears to have naturally been a very territorial beast and had been at first considered hard to domesticate, with only experts able to tame them. However, later in time we know that they had been domesticated and used as riding animals and pet companions. For sure, however, this isn't the kind of mudhorn that Mando runs into!

Looming large in the legend of the mudhorn, of course, were their eggs. In a most unlikely evolutionary quirk, it appears that the mudhorns of Arvala-7 laid fur-covered eggs that were edible by Jawas. On planet

Earth, only five species of mammal lay eggs and feed milk to their babies. In the scientific world, creatures who exhibit this eggs and milk business are known as monotremes; in contrast, the two other types of mammals, placentals and marsupials, reproduce through live births.

A Little Like the Mudhorn

The five species that share this extraordinary egg-laying trait are that most curious of creatures, the duck-billed platypus, and four echidna species (sometimes known as spiny anteaters): the Western long-beaked echidna, eastern long-beaked echidna, short-beaked echidna, and Sir David's long-beaked echidna, named in honor of naturalist Sir David Attenborough.

On Earth, these monotremes are found only in New Guinea or Australia. All monotremes are quite elusive creatures, and consequently little is really known of their daily routines and mating rituals. What we *do* know is that the echidnas use their fur as camouflage and spend much of the day hiding in empty burrows or fallen trees. The majority of their activity is at night, when they root out ants, termites, and other small invertebrates with their acute sense of smell. The natural habitats of the platypus, once described by Robin Williams as evidence that God gets occasionally stoned, and by Stephen Colbert as a creature that was still in its beta version, are rivers and waterways. This platypus is also nocturnal. They can spend more than ten hours a night hunting for their food, which comprises mostly shrimp and crayfish.

Planet Earth's Mudhorn

We have to go back almost three million years to find Earth's most convincing version of a mudhorn. Think turbo rhino. Think beefy unicorn. Think of a beast that, like the mudhorn with Mando, fought the greatest predators of its time, including humans. This creature possessed a weapon of immense power, but scientists are still trying to figure out how it worked. We are talking about the superbly named elasmotherium.

Elasmotherium, also known as the Siberian Unicorn or the Steppe Rhinoceros, was a genus of large rhinoceros that originated around 2.6

million years ago during the Late Pliocene epoch in what is now Asia. The fact that elasmotherium lasted until roughly forty thousand years ago means that they were alive at the same time as humans, so it's entirely possible that some readers of this book have ancestors that may have eaten elasmotherium, while others may have forebears that were mauled by it.

Scientists reconstruct elasmotherium as a woolly animal. This is mostly based on the woolliness typical in contemporary megafauna such as mammoths and the woolly rhino. Elasmotherium is also sometimes depicted as bare-skinned, somewhat like modern rhinos. The known specimens of *Elasmotherium sibiricum* grew up to fifteen feet in length, with shoulder heights reaching just over eight feet. Meanwhile, *Elasmotherium caucasicum* had a body length of at least sixteen feet, with an estimated mass of between four and five and a half tons. These stunning stats made elasmotherium the largest rhino of its day, comparable to the woolly mammoth and larger than the contemporary woolly rhino. *The Mandalorian*'s mudhorn appears to be based on quite a beast!

Introducing the Elasmotherium

In the domain of extinct mammals, the elasmotherium has grown into something of a celebrity. And the chief reason for its celebrity status is that (allegedly) huge horn. Some scholars say the elasmotherium horn was as long as nine feet. You'd not want to be skewered by *that* sucker.

To be doubly sure you get the idea of how titanic the horn of the elasmotherium was, consider this. The largest horn among living rhinos belongs to the white rhino. While the elasmotherium is impressively hung, the white rhino can only boast a horn thirty-five inches long on average. Yet, as with most theories in science, there's more to the elasmotherium horn theory than meets the eye.

Some scholars have recently suggested that the horn may in fact have been a much smaller bump affair. You can only imagine the controversy this generated, as we are talking of a beast whose legend is dominated by that huge, hardened spike. Moreover, there is also an ancient cave painting that is considered to be a person's rendering of an elasmotherium. Assuming the illustration is to be trusted, as we can't exactly email Stone

Age humans, this rendering shows a horn that is not as large as scholars have estimated it to be.

Introducing the Horn

Why do scientists believe that elasmotherium had such a huge horn in the first place? In short, archeological evidence. The outer layer of the horn probably consisted of keratin, the same stuff that makes up your fingernails and hair. In addition to this outer layer of keratin, archeologists think that the hard core of the horn was fortified by calcium and melanin deposits. This would bolster the horn's integrity and help prevent breakages. The trouble is, keratin doesn't usually fossilize, so these conclusions are based on a dome on the elasmotherium forehead that appears to be the base of the horn generating tissue. Coupled with that is evidence of a spine that appears to have been capable of supporting a large muscular hump. Such musculature is only truly needed if the creature is wielding a heavy object, so a big horn makes sense.

Living with the Horn

That horn wasn't just a fancy adornment either. Just look at the company this creature, the elasmotherium, was keeping. They lived alongside saber-toothed cats and cave bears, hyenas and lions, and that most cunning and deadly of animals—humans. Thus, we can only imagine that the elasmotherium had to be able to fight its way out of chance encounters with such predators. And it's not just the adult elasmotherium. Such a gargantuan beast might well be safe from harm, but there would be a constant necessity to protect their young.

Despite the majority of illustrations of the elasmotherium, scientists aren't 100 percent sure how big the horn really was. Even with a smaller horn, these beasts, like the mudhorn, would have been deadly. The elasmotherium most likely weighed in at around five tons. That's only slightly smaller than the largest living land mammal, the African bush elephant.

The legs of elasmotherium were longer than those in modern hippos, for example. Indeed, they were much more horse-like, which meant they could likely gallop a little like a pony. However, given the weight of an elasmotherium, they were more draft horse than thoroughbred, so probably didn't tear along with the grace of Gandalf's Shadowfax.

Colossal Horn but Named for Its Dentures

The horse and ancient rhino comparison might appear a little odd, but these two creatures are in fact quite closely related. Both beasts are members of the order known as parisodactyla, which means both are odd-toed ungulates, or they walked on just the one or three toes on each foot like their equine cousins. Elasmotherium were grazers. They fed on grass, though with teeth that were adapted to thicker, coarser grasses than the teeth of today's horses.

Elasmotherium had "hypsodont" teeth. This kind of tooth design involves tall crowns and enamel, which are folded into sheets, protecting the teeth from wearing away, and making chewing much more efficient. It's this feature of their teeth, and *not* their horn, which gives elasmotherium its name; *Elasmos* is ancient Greek for laminated, and *therion* means beast. They are hardy animals. Their molars never stopped growing, which prevented them from wearing away and causing elasmotherium to starve to death. (It's no surprise that the writers on *The Mandalorian* went for the name of mudhorn, rather than the less convincing mud-tooth.)

Like Tyrannosaurus rex, whose ancestors some scientists believe were to be found in China, elasmotherium likely originated in China, and gradually wandered westward to the Balkans. Modern-day rhinos have their horns on their nose, while elasmotherium had their horns right above their eyes, from which fact the nickname Siberian Unicorn derives.

You, the "IKEA" Monkey, and Elasmotherium

Remember the "IKEA monkey?" The creature so dubbed was Darwin, a Japanese macaque. At around six months old, Darwin decided he was bored so he escaped from his cramped crate, unlocked his owner's car door, and promptly walked out into the IKEA parking lot in North York on December 9, 2012. The rarity of the sighting of such a self-possessed monkey inspired headlines across the globe, making Darwin famous for five minutes.

We humans are more closely related to Darwin, the macaque, not the evolutionist who suggested such relations might exist, than rhinos and elasmotherium are to each other. Their most recent common ancestor lived roughly forty-five million years ago, which gives the two groups plenty of time to evolve into very different creatures. We're still not even sure if the elasmotherium skin was thick and rugged like a modern rhino, or if they were woolly like a mammoth.

Looks aside, elasmotherium was a very successful beast. When at their peak, they were to be found all the way from East Asia over to Eastern Europe. The last living species, *Elasmotherium sibiricum*, was given its name because a Russian princess, Ekaterina Dashkova, donated the first known fossils of the species to Moscow University in the early 1800s.

Elasmotherium sibiricum outlived other elasmotherium species, but finally fell victim to a changing climate. The lush grasslands, on which they used to graze, began to freeze at the beginning of the last Ice Age, and by around forty thousand years ago, the grasslands had become tundra, in a series of extinction events that also witnessed the end of the woolly mammoth, Irish elk, and saber-toothed cat.

This gargantuan beast, Earth's answer to the mudhorn, met the same fate as so many species in evolutionary history. And, to highlight just how gargantuan was *Elasmotherium sibiricum,* the largest living land mammal in Eurasia today is the robust European bison, weighing only around a tenth of what an elasmotherium did, which gifts us a glimpse of how these gargantuan prehistoric beings might have lived and behaved.

The answer to the question "did your ancestors munch a mudhorn?" is a very tentative "possibly!" The Siberian Unicorn lived alongside anatomically modern humans as well as Neanderthals, and it's not such an outrageous proposition to suggest that ancient hominids may have preyed upon these gargantuan creatures. After all, we know that early humans, in the likely form of *Homo erectus*, were hunting rhinos in what is now called the Philippines around seven hundred thousand years ago.

WHAT LIES UNDER MANDALORIAN SANDS?

(In which the author records his evolving thoughts, at the start of season three of *The Mandalorian*, about the state of the planet Mandalore)

Mandalore, the home-world of the Mandalorians, was a planet located in the Outer Rim Territories. Years of war left the planet inhospitable, forcing the Mandalorians to live within domed cities. The planet was heavily bombed in the Night of a Thousand Tears during the Great Purge of Mandalore, and millions of Mandalorians were killed. After the Purge, the Mandalorians were scattered in the galaxy, and some believed the planet was cursed.
—Mandalore, *Star Wars* canon (2023)

The Sands of Time

In anticipation of watching the season three opener of *The Mandalorian*, it strikes me that desert planets populate science fictional space. The legendary 1956 movie, *Forbidden Planet*, featured a starship surrounded by a huge, painted cyclorama featuring the desert landscape of *Altair IV*. Since then, desert planets have become a common setting in sci-fi. Frank Herbert's classic 1965 novel, *Dune*, famously adapted for the movies by David Lynch in 1984 and Denis Villeneuve in 2021, as well as a television miniseries by John Harrison in 2000, revolves around the habitat of the desert planet Arrakis (the so-called Dune planet of the book's title). The Dune franchise took its inspiration from Mexico and the Middle East, especially the Persian Gulf and the Arabian Peninsula.

And *Dune* in turn, of course, inspired the desert planets which prominently appear in the Star Wars franchise, including worlds such as Tatooine, Jakku, and Geonosis. *Collider*, the entertainment website and digital video production company, recently did a calculation to figure out how much of *Star Wars* happens on desert planets. Their numbers showed a collective runtime of 168,552 seconds, 45,177 of which are desert planet seconds. That's a dry and dusty 12 hours and 33 minutes, meaning desert planets make up 26.8 percent of the Star Wars franchise to date—a little over a quarter. The figure is similar for *The Mandalorian*, with a total "sand time" of 3 hours, 13 minutes, and 33 seconds in the first two seasons, a desert percentage of 19.99 percent.

The fictional obsession with desert planets is also somewhat reflected in fact. A 2011 scholarly study, "Alien Life More Likely on 'Dune' Planets" by Charles Q. Choi in *Astrobiology Magazine*, theorized that not only are life-sustaining desert exoplanets possible, but that they also might be more common than Earth-like planets. The study's models suggested that desert planets had a much larger and more tolerant habitable zone than ocean planets like Earth. The study also went on to suggest that Venus may have once been a habitable desert planet, around one billion years back, and that Earth will become a desert planet within a billion years due to the Sun's increasing luminosity. Happy days.

Desert

What lies under the surface of a blasted planet? We may soon find this out in season three of *The Mandalorian*. Mandalore was not a desert planet in the way that Arrakis and Tatooine were. On the contrary, Mandalore is described as having had a rich natural landscape, mostly unspoiled due to its sparse population. We are also told that this world was blanketed in lush forests, dense jungle, rolling hills and grasslands well suited to farming, inhospitable deserts of white sand, and many rivers, lakes, and seas. But war turned the world into a desert. Those that survived the devastation were resigned to do the same as the inhabitants of the planet Gallifrey on *Doctor Who*—engineer their own huge, protective dome in which to shelter.

We might be able to frame an answer to this question, of what lies under the surface of a blasted planet, by considering what is hidden under the sands of planet Earth's Sahara. Sure, the Sahara doesn't make up all of planet Earth. But with an area of 3,600,000 square miles, occupying almost one third of the entire African continent, it is the largest hot desert on Earth, and the third-largest desert overall, smaller only than the deserts of Antarctica and the northern Arctic, so it should act as a pretty good guide to the extensive deserted regions of Mandalore.

Saharan Sands

It's not hard to conjure up images of the Sahara in your mind. Limitless sandy landscapes cast in gold, a scorching Sun high above, and uninhabited, lifeless dunes and plains. Just the briefest glimpse of such a desert world may be enough to convince the uneducated observer that such a landscape is timeless. But geology knows better. Something must lurk under those many billions of tons of sand, just as they might on Mandalore, so let's find out what lies beneath the Saharan dunes.

The very name Sahara derives from a consonant Arabic word that literally means desert. Occupying the territory of ten different African countries at once, the Sahara is washed by the Atlantic Ocean and the Red and Mediterranean seas. Year after year, the Sahara grows. Year after year, the desert claims more and more space for itself. It expands to the south at an annual advance of three to six miles. We can only imagine that, on a blasted planet like Mandalore, such creeping desertification would also be a problem.

The Sahara Desert has been described as a cancer, and the African people are engaged in a titanic battle to stop the Sahara from spreading across Africa. They are planting a Great Green Wall of trees ten miles wide and 4,350 miles long, bisecting a dozen countries from Senegal in the west to Djibouti in the east.

Life on Mandalore

Given that so little has been said in seasons one and two of *The Mandalorian* about life on Mandalore, maybe we can better imagine this blasted alien world by considering the Saharan climate. For many, of course, the desert is synonymous with tall sand dunes of striking shapes, which scholars have named crescentic, linear, star, dome, and parabolic. Yet the sands occupy only 20 percent of the territory of the Sahara. No desert is sand alone. The main landscape is rocky plateaus. While the desert has an almost complete absence of vegetation and animal life, the rocky region has around four thousand species, among which there are hamsters, jerboas, antelopes, jackals, dune cats, and mongooses. It would be tempting to imagine similar species on Mandalore (the idea of a Mandalorian mongoose is very seductive), but, naturally, evolution would have taken a very different path on an alien world.

Most of these critters come out at night, as a nocturnal habit makes sense in the face of such a scorching daytime Sun. The temperature regime of the desert is also prone to large fluctuations during the day. The mean air temperature reaches 95°F, but the Sun heats the sand twice as much, so the conditions create a feeling of being in some huge and diabolical frying pan.

How hot is Mandalore? The hottest air temperature recorded in the Sahara is 136°F. And yet at night it can get as cold as 32°F, because the desert cools down at night as quickly as it heats up during the day. Among the desert's mountain plateaus, the story is even more extreme, with nighttime frosts and temperatures dropping below zero.

Rain is rare. In the central part of the Sahara, an average of 3 inches of rain fall per year. Yet, morning dew is not uncommon, and in the years 1879 and 2012, snow fell. It melted instantly. Despite the aridity of the Sahara, lasting lakes survive, while the only river flowing through the borders of this great desert is the mighty Nile.

Season Three Begins!

The Sahara is made not just of sand, but also of glass. The desert harbors areas where the dunes consist of large pieces of weathered glass, as well as the usual grains of sand. Scholars think that millions of years ago, in such places, large meteorites fell to Earth with a concomitant huge release of thermal energy.

Can we also expect glass on Mandalore? It's March 2023 as I write this, and season three of *The Mandalorian* has just begun. In "Chapter 17: The Apostate," Mando asks how he may atone for the sin of taking off his helmet. He is told that, according to Creed, "one may only be redeemed in the Living Waters beneath the mines of Mandalore." Mando replies, "But the mines have all been destroyed." The camera cuts to scenes of destruction, and we are witness to the atrocity of multiple mushrooming thermonuclear explosions as they litter the landscape of a doomed Mandalore.

With mushroom clouds, we can expect glass. Consider the early history of nuclear bombs on planet Earth. It used to be said that the energy released in the fission of a single nucleus of uranium is sufficient to make a grain of sand visibly jump. But would detonating a nuclear bomb in a desert truly turn the sand into glass? Yes, at least a portion of it.

Consider the New Mexico desert in 1945. After years of clandestine research and development, Manhattan Project scientists tested their first nuclear weapon, a plutonium bomb nicknamed Trinity, on July 16 of that year. The bomb was detonated over the New Mexico desert with a force of twenty-one thousand tons of TNT. The attendant fireball peaked at temperatures around fourteen thousand degrees Fahrenheit, far hotter than the surface of the Sun.

Trinity's mushroom cloud of devastation rose up more than seven miles into the sky. When the conflagration cleared, small lumps of radioactive green glass were found littering the ground for hundreds of feet around the blast site. And on doomed Mandalore, we had multiple blast sites. Under the impact of such devastation, the ground is transformed. The sand that had blanketed the desert the day before had been replaced by new and alien-looking, green-tinged lumps of material. These lumps of glass were later called trinitite, after the Trinity bomb, though the glass is sometimes called Alamogordo glass, as the bomb was blown not too far from the city of Alamogordo, New Mexico.

At first, scientists thought that trinitite was made when the heat of the fireball liquefied the sand on the ground. But a 2005 study in the *Journal of Environmental Radioactivity* conducted tests on trinitite samples, and scholars at the Los Alamos National Laboratory concluded that the chunks of trinitite were created when sand got scooped up into the atomic fireball, liquefied in the intense Sun-like heat, then fell back down to the ground and cooled. We can expect the same on Mandalore.

Glass on Mandalore!

Indeed, on writing this, imagine my delight in catching that first episode of season three, "Chapter 17: The Apostate." We see Mando and Grogu enter the workshop of the Armorer, when the following conversation occurs:

> **Armorer:** "You have removed your helmet. What's worse, you did so of your own free will. You are no longer Mandalorian."
> **Mando:** "The Creed teaches us of redemption."
> **Armorer:** "Redemption is no longer possible since the destruction of our home-world."
> **Mando:** "But what if the mines of Mandalore still exist?"
> **Armorer:** "All was destroyed in the Purge."

We then see Mando conjure up a relic from his person and place it on a table. Lo and behold, the relic is made of green-tinged glass! It may not be trinitite, but it could certainly qualify as a Mandalorian equivalent. Grogu coos and the conversation continues:

> **Armorer:** "Where did you come upon this?"
> **Mando:** "Jawas. They came upon by trade from a traveler who claimed to have visited the surface of Mandalore."
> **Armorer:** "This relic only proves that Mandalore's entire surface has been crystallized by fusion rays."

Bingo.

My personal excitement over this green glass prediction doesn't end there. Not only is episode two called "Chapter 18: The Mines of Mandalore," but we also get to see the effects of the planet's surface being crystallized by fusion rays. Mando and Grogu head to Mandalore with R5-D4 in tow, whose job it is to scan the possibly toxic planet. When we venture under the surface of Mandalore, we see that the destruction is so profound that the jagged and green crystallized rock is everywhere. The subterranean scenes are tinged in green. And Mando is soon attacked by Morlock-like Alamites who wield weapons carved out of the green glass, showing that they have adapted to Mandalore's new surface.

Rebuilding Mandalore

Prior to watching "Chapter 18: The Mines of Mandalore," I had done some research in an attempt to speculate whether the sands of the Sahara harbor any more secrets, perhaps something that might be key to mending and rebuilding a blasted Mandalore. This is what I found.

To uncover the most crucial clues of the Sahara, one needs to dig down and make a subterranean exploration. Deep in the very heart of this desert is a treasure under the dunes. While there is very little water on the Saharan surface due to irregular rainfall, under the sands of the Sahara are vast reservoirs of groundwater (the word engineers use to describe rainfall that has infiltrated the soil below the surface and collected in subterranean spaces). It is due to these pools that oases can be found in the Sahara, areas rich in vegetation.

Most of these oases arise in places where water from subterranean rivers finds its way to the surface. Some countries situated on the territory of Saharan groundwater are engaged in its extraction. The most lucrative land for such extraction is in the northeast of the desert, where Egypt, Chad, Sudan, and Libya are located. An aquifer is a body of rock that holds groundwater. And the aquifer in Libya has been used to extract groundwater since the 1970s. In 1983, the work began in earnest. A huge project was undertaken to deliver drinking water to the dehydrated settlements of Libya, and regular water supplies for all major cities of the country were established by 1996. This magnificent system is known as the Great Man-made River, and it delivers 6.5 million cubic meters of drinking water every day. In 2008, it entered the Guinness Book of Records as the world's largest irrigation project, consisting of over one thousand wells greater than five hundred meters deep, as well as countless pipes and reservoirs, all made possible by the Sahara's subterranean aquifer.

Rebuilding Mandalore

After surviving an attack by the Morlock-like Alamites in "Chapter 18: The Mines of Mandalore," Mando gets Grogu and they head deeper underground. They soon come to a precipice and gaze down upon the glorious ruins of the Civic Center of the buried Mandalorian city of Sundari. Looking like a submerged New York City, with "ground-scrapers" rather than skyscrapers, the very sight of Sundari is a symphony in cubism. Elsewhere in canon, we are told that the "New Mandalorians" built dome-shaped cities after centuries of war reduced Mandalore's surface to barren desert.

With Mando using his jetpack and Grogu tucked snugly into his cradle, our dynamic duo descends into the viscera of the Civic Center. As they drop down slowly into Sundari, we see that this is a gargantuan, multilevel city, with towering and translucent ground-scrapers in shimmering permacrete and transparisteel. *The Mandalorian* production team did an excellent job in realizing what we know from canon; that such an architectural style made Mandalore famous. We get mere glimpses of translucent or even transparent windows and panels, often set out in geometric designs, making possible vistas up and down through adjoining buildings and other structures, as much as through any window.

The design of Sundari calls to mind the famous 1924 science fiction novel, *We*, by Russian writer Yevgeny Zamyatin. Hugely influential on George Orwell's more famous dystopian 1948 novel, *Nineteen Eighty-Four*, Zamyatin's future glass city of OneState reflected the story's totalitarian tale of social uniformity and total surveillance. Zamyatin's glass city is a transparent hive, whereas Sundari is a city of necessity, the prismatic geometric blocks of the city itself and the encircling dome that surrounded it a monument to the Mandalorians' will to survive.

Mandalore's Great Man-made River

Domed cities have been a fixture of futurism since the early twentieth century and already appeared in sci-fi during the last quarter of the nineteenth century. Often inspirations for possible utopias, and situated on Earth, the Moon, or beyond, writers used domed cities as remedies for many ills: underwater glass-domed cities to rival Atlantis, or land-based builds meeting challenges of air pollution or nuclear devastation. The domed city has been used to represent the last stand of a dead or dying human race, and so the domed world becomes a symbolic womb that nourishes and protects humanity.

What's fit for a future Earth also looks fine for Mandalore. Sundari placed limits on its "imprisoned" inhabitants, but only within the context of the chaos of the world outside the dome. Given what we know about the great man-made river under the Sahara, perhaps the Living Waters, the ceremonial ground hidden in the Mines of Mandalore under the Civic Center, is also an aquifer. It certainly makes a lot of sense that the Mandalorians would build a city over such a resource, especially since according to ancient folklore the mines were once a Mythosaur lair, and the Living Waters were once used as a ceremonial ground where the royal family would bathe.

PART II
SPACE TRAVEL

HOW WAS TYTHON'S
FORCE-HENGE BUILT?

A trilithon is a structure made up of two large vertical stones supporting a third stone set horizontally across the top. It is commonly used in the context of megalithic monuments. The most famous trilithons are those of Stonehenge in England. At midwinter, Stone Age humans would gather to watch the Sun set between the Great Trilithon, a kind of seeing stone.

—English Heritage, *Understanding Stonehenge* (2023)

Force-henge

Mando is on a mission to conjure Jedi. In "Chapter 14: The Tragedy," he arrives with Grogu on Tython, a green and pleasant, yet uninhabited, rocky planet situated in the galaxy's Deep Core. In "Chapter 13: The Jedi," Mando had been told by Jedi Ahsoka Tano that planet Tython held the ancient ruins of a temple with a strong connection to the Force. Mando finds the Tythonian mountain on which sits the site of the Force-henge temple and places Grogu on the seeing stone. Grogu enters a meditative state while perched on the stone as battle soon rages about him. We watch as the dome-shaped seeing stone, ancient runes running around its base, seems to send a beacon beyond this green world. We wonder if the shimmering blue shafts of energy will soon summon any nearby Force-sensitives like Grogu.

The Force-henge looks fascinating. On close examination, the formation of Force-henge appears to be a mashup of a number of different traditions in terrestrial archeology. First, the henge is clearly derived from the kind of circle of standing stones typical of megalithic architectures of the ancient monuments of late prehistoric northwest Europe. The best example for comparison would be Stonehenge, as we shall soon see. Second, Force-henge also borrows from the long folkloric pagan tradition of places of spiritual mediation between this world and other worlds. This is, after all, kind of what Grogu is doing on that seeing stone. Third, when we consider the angle of the six main Force-henge stones, also known as the orthostats, which lean but are propped up, this also suggests that the production team included in their archeological mashup the so-called dolmen tomb tradition. A dolmen, or portal tomb, is a type of single-chamber megalithic tomb, usually comprising two or more upright megaliths supporting a large flat horizontal capstone or "table." Most date from the early New Stone Age and were sometimes covered with dirt or smaller stones to form a burial mound.

Next in the mashup, it's Middle Earth meets *Mandalorian*. The central stone of the Force-henge, the half dome-shaped seeing stone, is almost identical to Tolkien's Stone of Erech. The Stone of Erech was a great black stone, also spherical in shape and roughly six feet in diameter. It was set upon the hill of Erech in the land of Gondor and local hill tribes were made to swear an oath of loyalty on the stone. Tolkien's Stone became a mysterious and eerie place, shunned by the people of the valley, who claimed it had fallen from the sky and was haunted by restless spirits. It's also where Aragorn rides to call the dead to honor their oath to ride out and defend Gondor.

The Force-henge mashup has a few more derivations from our archeological past. The seeing stone has a banded inscription of runes across its equator, so it's also essentially a rune stone. This makes it comparable to the Viking-age monuments of Scandinavia. However, the location of the Force-henge is quite unlike that of any European prehistoric architectures or rune stones. The Force-henge temple is hilltop, perhaps inspired by non-European, such as Buddhist, traditions which favor hilltop-shrine topography. Finally, Boba Fett actually refers to the temple as a "henge," so let's make our main point of reference between Force-henge and Stonehenge.

Imagine we freeze-frame a temple scene from "Chapter 14." We time our image capture just at the point where we can clearly see Mando walking across the Force-henge "floor." Once we have our image, we then make an estimate of how tall Mando appears to be on our picture. We do this so we can measure how many Mandos it takes to define the dimensions of the Force-henge. Assuming Mando to be six feet tall (remembering it's not Pedro Pascal in the suit but John Wayne's grandson, Brendan Wayne, who is six foot in stocking feet), the Force-henge turns out to be roughly six Mandos high and sixteen Mandos in diameter. That's around thirty-six feet high and ninety-six feet across.

Stonehenge

How does that compare with planet Earth's most famous henge? Stonehenge is one of the most famous and recognizable historical sites on planet Earth. But for all its fame, it remains a deeply mysterious place, an enigma we know very little about. What we do know is that Stonehenge is a very complex prehistoric monument on Salisbury plain in Wiltshire, England. It is a World Heritage Site. It's Britain's wonder of the world. The monument was built at the dawning of a new age. In Egypt, the Pharaohs would soon start to build pyramids. In Britain, metals technology had only just been introduced. Centuries later, tin and copper would be mixed to form bronze. With the discovery of alloys, the British Bronze Age would start, and the Stone Age would eventually become a thing of the past. But for now, stone is all.

Do yourself a favor. Get on Google and take a virtual tour of Stonehenge. Better still, if you can, visit the site in person. Perhaps you will feel some of the awe that Grogu feels when he sits upon the seeing stone. When you become a regular visitor to the site, no matter how many times you visit, the monument never loses its sense of magic. It's positioned in the middle of the wide-open Salisbury plain, raised up and looking out at a wonderful 360° view, just like the temple of Tython.

The stones themselves have remained upright since they were erected by our Stone Age ancestors roughly five thousand years ago. The site is a place to come and contemplate the passing of time, how small and irrelevant our passing lives are, compared to the generations of humans that these stones have seen, or to merely marvel at its construction.

Stonehenge's construction began about five thousand years ago. The stones that most people associate with Stonehenge, the circle of sarsen stones at the center of the monument, were erected first. These sarsen stones run up to roughly twenty-three feet high, with the circle itself around ninety-seven feet across. Quite similar in size to the Force-henge—and the similarities don't end there.

Tython is said to have had a pivotal role in the history of the Jedi Order. A lush world, hugely rich in the Force, the planet eventually became uninhabited when the ever-changing hyperspace routes in the region wandered away from the planet and Tython became masked in myth. One such myth, held by some Jedi scholars, was that the planet was the site of the Jedi Order's first temple. Another was that this legendary planet was filled with Jedi wisdom since time immemorial, and if a keen student were to uncover it, they would be tapping into unparalleled power.

Stone Temple

Stonehenge too is a temple. It's a prehistoric monument unparalleled in terms of its profound heritage power, and it still holds secrets that might tell us more about how we humans used to live. The sarsen stones weren't just thrust up any old how. They were carefully aligned with the movements of the Sun. Just as the shimmering shafts of energy draw down on Grogu when he sits on the seeing stone, Stonehenge draws solar energy into the center of its circle.

Imagine Grogu comes sightseeing on planet Earth. He's interested in the Earth's prehistoric monuments and visits Stonehenge. If he sits in the middle of the sarsen circle on Midsummer Day, the longest day of the year, he would see the Sun rise just to the left of the Heel Stone, an outlying stone to the northeast of the monument. This Heel Stone may once have had a partner stone. Archaeology scholars have found a large hole to the left of the Heel Stone, which may have held a partner, the two stones together framing the sunrise.

Should Grogu have been so impressed that he again visited Stonehenge, this time on Midwinter Day, the shortest day of the year, there'd be another tale to tell about Stonehenge as a temple. In winter, Grogu would have to turn through 180° from his Midsummer position and face southwest. Here he would see that, originally, the Sun would have set between the two uprights of the tallest trilithon at the head of the sarsen horseshoe. The Sun would have dropped down into the Altar Stone, a sandstone megalith placed across the solstice axis. These days, the effect has been somewhat lost due to the fact that half of the trilithon has fallen at some point in Stonehenge's long history. Scholars have also found, using laser power to pierce the outer layers, that those stones framing the solstice axis were the most carefully worked and shaped using hammerstones, creating vertical sides that framed the movement of the Sun.

In what way is Stonehenge a temple like the one on Tython? Well, first consider Stonehenge's axis of construction. The entire layout of the temple is oriented in relation to the solstices. The solstices are the two times of the year when the Sun is farthest away from the equator, so they represent the extreme limits of the Sun's apparent movement across the sky (the word solstice comes from the Latin world *sol*, meaning "Sun," and *sistere*, meaning "to stand still"). At Stonehenge, the axis of the solstice is also marked by the so-called Station Stones. These are placed in a rectangle on the edge of the surrounding circular ditch, with the short sides of the rectangle on the same alignment as the sarsen stones.

Stonehenge's "Avenue" links the main part of the monument to the nearby River Avon. The Avenue is an ancient avenue made up of parallel banks and ditches. The Avenue is also crucially linked to the movements of the Sun. Its final straight stretch close to Stonehenge is aligned on the northeast to southwest solar axis that Grogu observed on the solstices. Recent archeological excavations along the Avenue have found some fascinating discoveries. The earthworks of the ancient Stonehenge engineers seem to have followed the line of some periglacial stripes (natural ridges and gullies) that were already there. These stripes are natural features from glaciation, but the presence of periglacial stripes that just happened to line up with the solstice may have been noticed by Neolithic engineers, which could explain why they built Stonehenge on this particular site.

Temple to the Sky

Like the Jedi presence on Tython, the deeper meaning of Stonehenge has been shrouded in mystery. Marking out the movements of the Sun must have been a crucial design element for our Stonehenge engineers, since they clearly exerted such enormous effort to build and carefully align the monument. On average the sarsens weigh twenty-five tons, with the largest stone, the Heel Stone, weighing about thirty tons. These stones were brought from the Marlborough Downs, about twenty miles away. And if you think that's no mean feat, consider Stonehenge's bluestone monoliths. Bluestone is the name given to the somewhat smaller stones at Stonehenge. They weigh between two and five tons each. Although they don't look blue, they have a bluish blush when freshly broken or wet. But here's the thing about those bluestone monoliths. They were transported to Stonehenge all the way from the Preseli Hills in southwest Wales, a distance of at least 180 miles. Yes, that's right. Thousands of years ago, somehow, the Stonehenge engineers, who were in fact just farmers, herders, and pastoralists, brought these two- to five-ton bluestones on a journey of almost two hundred miles. (For comparison, the distance between New York City and Boston is roughly 190 miles, as the crow flies, and the distance between New York City and Washington, DC, just over two hundred miles.) I live a mere one hundred miles from the Preseli Hills, in south*east* Wales. And today, even when equipped with my Land Rover and a flatbed trailer, without some kind of use of the Force, I'm pretty sure I would find it impossible to cart a bluestone home.

For generations, people have argued over Stonehenge, the most iconic and mysterious structure ever built in the British Isles; its origin, meaning, and purpose. But recently, over the last decade, a team of dedicated archeologists made an amazing discovery, one that has had us rewrite prehistory books.

Merlin and the Lost Circle of Stonehenge

Using the latest scientific techniques, archeologists unearthed the remains of another immense stone circle, dismantled just before

Stonehenge was created. This may well be the true origins of one of the world's most famous monuments. This new discovery is linked to the legend of Merlin, the Welsh wizard, a most Jedi of characters. There is an ancient myth, one first recorded in the Middle Ages, that Merlin led men far to the west of Britain, to Ireland, to the land of giants, where he found the stones and, Jedi-like, used some kind of magical force to transport them to Salisbury Plain. This myth is clearly fantastical, yet myths can contain grains of truth, which is the case with the lost circle of Stonehenge.

The bluestones originally formed an even more ancient long-lost monument. Archeological data suggests that the bluestones had been quarried almost four whole centuries before Stonehenge was built. Could the Stonehenge bluestone circle have originally stood in West Wales? After all, the Merlin myth tells a tale not just of the origin of the stones, but also of an existing stone circle in the west, a monument known as "Giants' Dance."

After months and years of searching, the site at Waun Mawn (Welsh for "peat moor") gave up its ancient secrets. When unearthed, the diameter of the Waun Mawn circle was revealed to be 110 meters, exactly the same as the outer perimeter around Stonehenge. The chances of the two stone circles having precisely the same dimensions are slim indeed. What's more, the archeologists found evidence of a direct connection between the two monuments. They used computer software to confirm that a pentagonal-shaped stone hole at Waun Mawn was a perfect fit for one of the stones at Stonehenge. It fitted like lock and key.

Next, they used a most incredible technique. This technique is able to detect ancient sunlight in the soil at Waun Mawn. How? By measuring the remnant energy in grains of quartz in the soil since it was last exposed to light! The results that came through showed a likely construction date of around 3300 BC for Waun Mawn. Another perfect fit, as it was "shortly" before Stonehenge. Part of Stonehenge's secret history had been unearthed and the lost circle had been found. Stonehenge was first built not on Salisbury Plain but in the Welsh hills centuries earlier by people who were steeped in a culture of megalithic architecture.

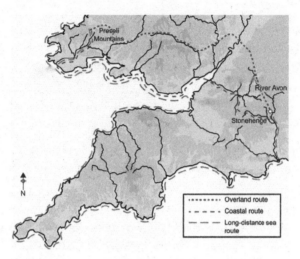

Figure 2. The possible routes of the bluestones to Stonehenge.

A Stone Age Thought Experiment

How do archeologists believe the Stonehenge engineers moved those bluestones from Waun Mawn to Salisbury Plain? In the Stone Age, there were no wheels, no axles, and no metal tools yet, at least not in ancient Britain. The countryside was covered by vast woods and forests. The mystery of the build still leaves many academics wondering how the feat was achieved.

Let's play a thought experiment. Maybe this experiment will help us imagine how an ancient Jedi culture assembled their Force-henge hilltop temple. Picture yourself in Stone Age Britain. No use of the Force is allowed, and there are no Force-sensitives like Grogu to help you in your task. Like many other humans in your area, you farm pulses, barley, and wheat, and you rely on wild food and resources. (Farming, maybe the most important development in human history, was first introduced to Britain around 4000 BC by people who brought farming techniques to the island by boat.) Rather than settle in one place, your tribe moves around territories, and these territories are focused on great communal monuments. Your local monument is Stonehenge.

With the primitive technology available to you, which route do you prefer to bring those bluestone monoliths to Stonehenge? Option one, the coastal route; option two, the long-distance sea route; or option three, the overland route. Let's look at each in turn so you can pick your eventual path.

The coastal route: This route leaves the Preseli Hills heading south to the coast where the huge stones were loaded onto boats, brought around the south "Wales" coast to "England," then up what's now known as the River Avon toward "Salisbury Plain." A sea route for the bluestones has been popular for decades. However, the position of the monolith quarry north of the Preseli Hills makes it unlikely that the bluestones were ever taken south to the sea. To do this, the Stonehenge engineers would have had to transport the monoliths up and down steep slopes to the ocean.

The long-distance sea route: There are similar problems with this longer and more hazardous sea route (again, initially along the south "Wales" coast, but then all the way around Land's End to the southern mouth of the Avon), mostly down to the dangerous cliffs and currents along this way. Furthermore, in 3000 BC, boat construction was in its infancy at best. The boats would not have been sturdy enough to hold multi-ton stones. It would have been a very risky prospect to maneuver a large megalith over water. It was tried in the year 2000, a millennium project, and the stone sank within half a mile of leaving land!

The overland route: Even though the shorter sea route is around 180 miles long, and the land route around 220 miles, the overland route is more favored. This route heads east along a series of conjoined flat-bottomed, glaciated valleys, then south to Salisbury Plain. Studies show that animal migration as a proxy for human routes of movement through these areas identified this overland route as a major corridor in prehistory. This thought experiment maybe has a solution that you'd not expect. The sea route *seems* easier, but a closer look tells a different tale.

A Force-henge Thought Experiment

What might the gathering of the bluestone monoliths and the building of Stonehenge tell us about the creation of the Force-henge temple on Tython? To answer this question, we naturally have to make many assumptions about *Star Wars* history. And it would be a rather boring solution if we merely suggested that Jedi and other Force-sensitives like Grogu simply performed a hive-mind use of the Force and floated the orthostats to the selected hill.

In *The Mandalorian* and the Star Wars Universe, some things are portrayed as supernatural such as the Force which, when used skillfully, can supersede our normal laws of nature. But, given that the laws of nature are well thought out, having stewed and simmered over thirteen-odd billion years of evolution, it is always wise for the *Star Wars* writers to wonder in what way the supernatural should be allowed to supersede, and in what situations.

A comparison with the Harry Potter Universe may help here. Both fictional Universes place limits on the fantastic feats to make the fiction more believable. J. K. Rowling has gone on record suggesting that, though not explicitly stated in her books, wizards could not simply conjure money out of thin air. An economic system based on such a possibility would be grimly flawed and highly inflationary. Perhaps that's also why a limit was placed on the use of the Sorcerer's Stone for alchemy. The Stone's abilities were described as extremely rare, possibly even unique, and possessed by an owner who did not exploit its powers. Dumbledore also says there is no spell that can bring people back from the dead. Sure, they can be reanimated into compliant beings on a living wizard's command, but they are little more than soulless zombies with no will of their own.

All of the above is my way of pointing out that there are limits to the way a writer can create a fantastic fictional Universe, just as there are limits to science. So, for our thought experiment, let's assume that Jedi can't simply conjure the orthostats up that hill in the same way that wizards should not simply be allowed to conjure money out of thin air.

Perhaps a more agreeable history for the Force-henge temple is this: We know that Tython goes far back in Jedi legend. And we know that Jedi artifacts and holocrons latently litter the planet. So, just maybe, early Jedi exponents slowly learned how to marshal a hive-mind use of the Force to build such temples. The rest was done with blood, sweat, and tears, just like Stonehenge.

Remember those scenes of Yoda training Luke on Dagobah in *The Empire Strikes Back*? We see Luke learning how to use the Force. Performing a handstand with Yoda atop one leg, Luke is seen being coached in the balancing of boulders into precarious piles. Meanwhile, Luke's X-Wing *Red Five* sinks entirely into the swamp. Yoda suggests Luke use the Force to retrieve *Red Five*. Luke is exasperated, saying to Yoda, "Master, moving stones around is one thing, but this is totally different."

Yoda's reply is very telling, "No! No different. Only different in your mind. You must unlearn what you have learned"; presumably a reference to how the gravitational field affects objects of different masses. We then see Luke trying to pull off the miraculous feat of using his mind to drag the X-Wing back out of the swamp. He fails.

"I can't, it's too big," Luke says.

"Size matters not," replies Yoda, who then shows Luke just what he means as he pulls off the feat of not only levitating the X-Wing back out of the swamp, but also floating it to a patch of relatively dry land.

Switch scene to ancient Tython where we have the prospect of members of the Jedi order, and maybe other Force-sensitives, engaged in the challenge of battling with the planet's gravitational field to move the orthostat stones into the configuration we see in "Chapter 14" of *The Mandalorian*. Now, it seems that it was relatively easy for Yoda to Force-lift the X-Wing, though an out-of-practice Obi-Wan seemed to struggle with the weight of a young Leia when he used the Force to correct her fall from a building in "Part II" of Disney's *Obi-Wan Kenobi* TV series, so it seems reasonable that there must surely be an upper limit on what a single Force-sensitive can lift. Size may matter not, as Yoda claims, but we know that use of the Force drains a person's energy.

We have no estimate of what a single orthostat would weigh, but consider Brú na Bóinne, otherwise known as the 5,200-year-old Newgrange passage tomb in Ireland's Boyne Valley. It is estimated to contain well over 220,000 tons of stone. That's a lot of stone. If the Tython Force-henge is anything like that, it would have taken an awful lot of collective energy to shift. But first, those Jedi minds have to be trained to overcome the prejudice of physics.

Inertia: Physical

In normal circumstances, what would be the biggest obstacle in moving one of those six megaliths from their original resting place to become a leaning and propped-up orthostat as part of the Force-henge? The answer is inertia.

Inertia is a curious phenomenon. It is, in some senses, the force that holds the Universe together. Inertia is a resistance to change and, in particular, resistance to changes in motion. We learn the principle of inertia early in life. Even as children, we know that it takes a force to get something going, to change its direction, or to stop it. The concept of mass is tied up with the concept of inertia. Mass is that quantity that is solely dependent upon the inertia of an object. The more inertia that an object has, the more mass it has. A more massive object has a greater tendency to resist changes in its state of motion.

Let's imagine we apply a steady force to Mando's *Razor Crest* spaceship. Don't worry too much about *how* we're doing this, just that we *are* doing it. When Mando's *Razor Crest* picks up speed, we say it accelerates. British physicist Isaac Newton was one of the first to realize the link between force and acceleration. Newton put it something like this: the acceleration of a body is in proportion to the force applied to it, and inversely proportional to the body's mass, which is also sometimes called the inertia of the object. Now, these last points are of interest. They mean that the bigger the force, the faster an object picks up speed, but also that the bigger the body mass, or inertia, the harder it is to get moving quickly (this makes sense, as it's easier to get a mini car going than a monster truck).

It was Albert Einstein who showed that mass is not constant. A body like the *Razor Crest* has a rest mass, the mass when it is stationary. But, as the speed of *Razor Crest* increases, its inertial mass also increases. For low speeds, those much less than light speed, this increase in mass is barely perceptible, but should *Razor Crest* get close to the speed of light, its mass starts to increase swiftly toward infinity. In theory, *Razor Crest*'s mass would become infinite if it accelerated all the way to light speed. But, because the acceleration of a body like *Razor Crest* in reaction to an applied force is inversely proportional to its inertial mass, as light speed is approached, the force needed to reach light speed also becomes infinite. That's why, according to *our* current physics, it's not normally possible to accelerate *Razor Crest* to the speed of light. In short, and in the spirit of Einstein, if you gave *Razor Crest* more and more energy, instead of going faster and faster, it just gets heavier and heavier. No matter how much of a shove you gave it, *Razor Crest* would still be going less than light speed.

Newton also had something to say about inertia. He enshrined inertia in his first law of motion when he said that an object tends to remain at rest, or in uniform motion in a straight line, unless compelled by other forces to change its motion. What does Newton mean?

Well, imagine an orthostat floating through space. Don't worry about how it got there, or who gave it such an almighty push that it is traveling at a steady speed through a planetary system. Big rocks, after all, are quite common in deep space. Newton's first law says that our orthostat would indeed keep speeding on forever—unless it got another shove. This may seem normal, the tendency of a mass to keep on going forever unless compelled not to. But, when you think about it, why *should* things keep on going forever? It's kind of weird! We take it for granted, due to our life's experience, but when we examine it, it can begin to feel like an "alien" phenomenon. And that phenomenon's name is inertia.

Psychological Inertia and the God Particle

As if the laws of physics weren't enough for the ancient Jedi to contend with, there's also the fact that (as Yoda implies with his advice of "you must unlearn what you have learned") inertia may also manifest in the minds of Jedi recruits. A rookie Jedi's intuitive sense of how inertia works would have enabled them to exercise a degree of control over the stones, at least in conventional ways. By applying external physical forces, they know that they can use blood, sweat, and tears to overcome the inertia that keeps the stones still in their original resting place, and shift the stones to where they want them, just as the Stonehenge engineers did. But, to make life "easier," what do more experienced Jedi teach the rookies about how they can overcome the inertia of the mind, that barrier that divides ordinary mortals from Jedi? This is all assuming, of course, that way back when, in the foggy mists of time on Tython, there was anything other than rookie Jedi to begin with.

Consider the God particle. Well, "God particle" is what we terrestrials call it. The particle's other name is the Higgs boson. It is the elementary particle associated with something known as the Higgs field, a field that gives mass to other elementary particles of the cosmos such as electrons and quarks. We know a particle's mass determines how much it resists changing its speed or position when it encounters a force, but not all elementary particles have mass. For instance, the photon, the particle of light, has no mass at all.

To make decent progress, the Jedi would have to contend with the Higgs boson. It gets its mass just like other particles: from its interactions with the Higgs field. The current belief is that there may be more than one Higgs boson. One theory of new terrestrial physics predicts that there may be as many as five Higgs bosons. While the Higgs boson gives mass to the quarks that make up a proton, it is only responsible for giving a proton about 10 percent of its mass. The rest of a proton's mass comes from the interactions of the quarks and other forces.

The story of particle mass starts straight after the Big Bang. During the first mere minutes of the cosmos, almost all particles were massless, traveling at the speed of light in what's sometimes called a "hot primordial soup." At some point during this time, the Higgs field turned on, permeating the cosmos and giving mass to the elementary particles.

The Higgs field changed history when it turned on. It changed the way in which particles behave. Scholars of the cosmos, cosmologists, use different metaphors to try explaining the Higgs field. Some compare the field to a vat of thick syrup, which slows some particles as they travel through. Others picture the Higgs field as a horde of paparazzi. As A-list celebrities pass through, such as Jon Favreau and Rosario Dawson, the paparazzi surround them, slowing them down. But less-known faces, such as Pedro Pascal without his helmet, pass through the paparazzi relatively unbothered. In this analogy, popularity is synonymous with mass; the more popular you are, the more you will interact with the paparazzi, and the more "massive" you will be. As to why the Higgs field turned on in the first place, and why some particles interact more with the field than others, we just don't know right now. Maybe the Jedi did.

Since the Higgs boson has the role of generating the mass of other particles, it may be the secret to partly switching off the phenomenon of inertia. The Jedi would have to learn some way of switching off the Higgs field. This is no trivial matter! It would be far too tall a task to do this everywhere, of course, which is rather fortunate for the cosmos, as inertia is the phenomenon that holds the Universe together. But if they could learn to switch off the Higgs field around an orthostat, say, then the particles in the stone would stop their interactions with the Higgs field.

Yet even that wouldn't be enough. Sure, the Higgs field gives mass to elementary particles, but as these still only account for a small proportion of the Universe's mass, the Jedi would also have to find a way of turning off the strong nuclear force. The protons and neutrons in their nuclei of atoms get almost all their mass from this strong nuclear force. Protons and neutrons are each made up of three quarks bound together by gluons, the particles that carry the strong force. The energy of this interaction between quarks and gluons is what gives protons and neutrons the rest of their mass.

The choice facing the Force-henge engineers would be twofold. One, copy the Stonehenge engineers and use blood, sweat, and tears to overcome the stones' inertia and use conventional physics to build the henge. Or, two, rewrite cosmic physics in the most revolutionary way, temporarily switching off both the Higgs field and the strong nuclear force for the stones, constructing the henge the "easy" way. As a final word on the way in which the Jedi evolved to become master exponents of the Force, and were able to supremely wield it, consider the evidence in Disney's *Obi-Wan Kenobi* TV series. In "Part V," we see Darth Vader use the Force to stop a transport ship taking off from Jabiim. Vader is also seen throwing medium-sized stones around in "Part VI," only to be later outclassed by Obi-wan himself who lifts practically an asteroid belt's worth of stones in the same battle and throws them at Vader. The same scene also has the wonderful line, uttered hesitantly by Obi-wan, "Goodbye . . . *Darth*."

HAS *THE MANDALORIAN* MADE A PARSEC MISTAKE LIKE THAT IN *A NEW HOPE*?

"The parsec is a unit of length used to measure the large distances to astronomical objects outside the Solar System. The parsec is approximately equal to 3.26 light-years or 206,000 astronomical units; in other words, 30.9 trillion kilometers."

—Mark Brake, *Alien Life Imagined* (2012)

"It took [Werner] Herzog less than one minute after meeting [Jon] Favreau to be drawn to the project, later saying of the conversation: 'I saw the Universe. I saw costumes. I saw the round horizon. I saw the spacecraft. I saw an entire Universe. And I knew this was really big.'"

—Geoff Boucher, "*The Mandalorian*: Werner Herzog Hails Disney+ Series As 'Cinema Back at Its Best'" (2019)

Oops

As f*ckups go, it's pretty famous. We all know the story. Han Solo boasts in *Star Wars: A New Hope* that his fictional starship the *Millennium Falcon* "made the Kessel Run in less than 12 parsecs." It's a little like bragging "I flew down Route 66 in less than a thousand miles" or "wow, it only took me three miles to boil this egg" or "damn, man, it's really cold today; reckon it's half past four." Okay, you get the picture. It makes no sense.

Star Wars fans got irritated by science pedants who spoiled the fiction with hard facts about the parsec. It's a measure of distance, not time. Perhaps the problem wasn't helped by the fact that one particular measure of *distance* in space is the light-*year*. When Han seemed to repeat his parsec proposal in *Star Wars: The Force Awakens*, astrophysicist Neil deGrasse Tyson upset some purists by suggesting that the franchise was "unashamed of inanity."

Tyson might be right. Why? Because *The Mandalorian* seems to have already made another, maybe smaller, mistake about the parsec. In "Chapter 1: The Mandalorian," Mando meets the mysterious, untrustworthy, and devious individual known as The Client. (The Client is played superbly by German film director, screenwriter, and actor Werner Herzog.) Only twenty-five minutes into this new *Star Wars* extravaganza, the parsec bomb is dropped. Mando walks into The Client's bunker and the following exchange occurs:

> **The Client:** "Greef Karga [the leader of the Bounty Hunters' Guild] said you were coming."
> **Mando:** "What else did he say?"
> **The Client:** "He said you were the best in the parsec."

Best in the Parsec

It's pretty clear that *The Mandalorian* is full of tropes from Western movies, so it doesn't seem too much of a stretch to suggest that "best in the parsec" is some kind of space cowboy version of the Western trope of "the fastest gun in the west." You know the kind of cliché. The fastest gun in the west was the ultimate quick draw. Nobody can keep up with this guy. Whether hero or villain, the fastest gun can draw and fire a gun slicker than practically anyone on the planet, or at least anyone in the good old Wild West. Moviegoers lapped this stuff up. Especially in the so-called Golden Age of spurs-and-saddles films between 1940 and 1960 when up to 140 Westerns were released *every* year.

By analogy, we can assume that The Client is paying Mando a compliment. Mando clearly comes with a reputation. Word has gotten around. He's not to be trifled with, not just on this planet, but in the entire parsec. And that's where the trouble starts. Here's the thing. Cosmically speaking, and as The Client should really know, a parsec really isn't very far.

The Parsec

Let's unravel what "parsec" means and where it comes from. The word parsec is a portmanteau of parallax and arcsecond. Parallax has to do with the measure of astronomical distances. Luckily, humanoids like The Client have two eyes. This gives us what's known as binocular vision. When The Client looks at a nearby object, his left eye sees the object at a slightly different angle than his right eye. His brain puts these two images together, compares them, does the geometry, and gives him a sense of distance to that object. Depth perception, if you like.

Why not try it for yourself? Keeping your fist closed, extend your middle finger upwards. Now, looking at your finger, as you blink one eye and then the other, you can see that your finger appears to shift position relative to more distant objects in the room, or wherever you happen to be. That shift is called parallax. The amount of shift will depend on how far apart your eyes are. So, Admiral Ackbar's shift would be far greater than Chewbacca's, for example. The amount of shift will also depend on how far away the object is.

If you know the distance between your eyes, a distance known as the baseline, you can apply some math and figure out how far away the object is. If the object is close by, it shifts a lot; if it's farther away, it shifts less. It works pretty well, though it does put a limit on how far away we can reasonably sense distance with just our eyes.

Stars are way beyond that limit. If we want to measure star distances using parallax, we need a much bigger baseline than the mere inches between our eyes (or an entire foot, in the case of Admiral Ackbar). Once astronomers worked out that the Earth went around the Sun (more about this in a moment), they soon understood that the Earth's orbit could act as an enormous baseline. So, if we look at a star when the Earth is at one point in its orbit, then wait six months for our planet to go around the Sun to the opposite side of its orbit and look at the star again, in theory we can determine the distance to that star . . . all assuming we know the size of the Earth's orbit. In itself, no mean feat! That's why, historically, knowing the length of the astronomical unit, the Earth-Sun distance, has been so important.

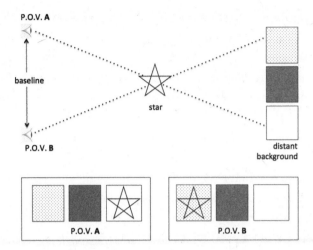

Figure 3. From point of view A, the star appears to have a blank background. But from point of view B, the background is dotted.

A Couple of Revolutionary Parallaxes

A most spectacular tale of the use of parallax involves the silver-nosed Danish astronomer Tycho Brahe. Heir to a number of Denmark's principal noble families, Tycho was well known in his lifetime as an astronomer. He had already, in 1572, made observations of a supernova in the constellation of Cassiopeia. This revolutionary discovery contradicted the ancient wisdom of Aristotle. The heavens were meant to be unchanging in Aristotle's view. Tycho's "new star" clearly showed a change in the sky, whereas previous consensus was that new stars were just some local phenomenon in the atmosphere. But Tycho used parallax. His exacting data showed that the supernova did not change positions with respect to the other stars. It had no parallax. So it couldn't have been local, as a nearby object would have a big parallax shift.

Come the great comet of 1577, Tycho once more used parallax in a revolutionary way. By measuring the parallax for the comet, Tycho was able to show that the comet was beyond the Moon. Again, this contradicted Aristotle's view, a view held for almost two thousand years, that comets, too, were atmospheric phenomena. As with the case of the supernova five years before, the great comet was a clear and obvious change in a sky that was meant to be unchanging. These examples show how the mere measurement of parallax can lead to revolutionary results, and are evidence of why many folks consider Tycho to be the greatest of all pre-telescope astronomers.

Back to the Parsec

Since parsec is a portmanteau of parallax and arcsecond, and we've already discussed parallax, let's consider the arcsecond. Have you ever been stargazing with family, friend, or droid? If so, you may know how frustrating it can be to point out a particular star from all the others, especially if the "seeing" is good and the sky dark. Because there are so many stars, it can be easy for others to get lost in your explanation.

Humans have been stargazing for as long as we've been smart and curious creatures. There exist star charts from various cultures, thousands of years old. Clearly, our ancestors also saw the need to be able to distinguish one star from another, just for the record. After all, stars don't change their apparent positions in the sky for thousands of years, so the maps hold true for some time.

The arcsecond helps with pointing out a particular star. Imagine you are stargazing with the family droid. The droid is programmed to help you appreciate the night sky and tell one star from another. "Picture the dome of the night sky as the face of a clock," the droid tells you in a voice that sounds somewhere between Jarvis, Stephen Hawking, and Optimus Prime. "And that clock is divided into hours, minutes, and seconds," the droid continues. "As with their use of a clock, you humans separate the dome of the sky into degrees, rather than hours, and each degree is composed of arc minutes and arc seconds. There are sixty arc minutes in each degree, and each arc minute is made up of sixty arcseconds."

"Just how big is an arcsecond?" you ask the droid.

"Consider the full Moon," they reply. "Your planet's satellite covers approximately one half of one degree of night sky. That's equivalent to thirty arc minutes, or 1,800 arcseconds." Thinking of the full Moon as being 1,800 arcseconds across just shows you how tiny an arcsecond is.

Putting together what we know about parallax and arcsecond, we can now define a parsec as the distance from the Sun to an astronomical object that has a parallax angle of one arcsecond. In other words, a parsec is the distance an object would have to be for astronomers to measure it and find that it appears to be 1,800 times smaller than the full Moon. You can imagine that it would take a huge distance for the object to be measured as only one mere arcsecond. The parsec distance is, get this, equivalent to 206,000 Earth-Sun distances (a.k.a. astronomical units.) That's approximately equal to 19 trillion miles; that's 19 million *million* miles.

Back to the Client

Putting all these findings together, we can now see what The Client is saying to Mando. In suggesting that Mando is the "best in the parsec," The Client is claiming that Mando is the best bounty hunter within a radius of at least 19 million million miles. That seems like an awesome compliment. But let's see what it would mean in practice if someone said it to Mando on planet Earth. What competition would his reputation have in the parsec? In short, none whatsoever. (Having said that, "best on Earth" isn't a bad boast, but the writers are trying to show off by using parsec instead, so the objection remains!) Using Earth as the center of an enormous circle, one parsec's radius wouldn't even be a big enough distance to reach the next star. The nearest star to Earth, Proxima Centauri (a member of the triple system of Alpha Centauri), has a parallax of only roughly 0.768, yielding a distance of 4.24 light-years, which is 1.302 parsecs.

The Client appears to have given Mando a backhanded compliment. We have to assume that this is unintentional, as we know The Client is trusting Mando enough to hire him for the job of finding Grogu, especially as The Client promises the lucrative reward to the bounty hunter of a container full of beskar. Yet, "best in the parsec" is a little like saying "I always feel more intelligent after reading your work" or "it's really difficult to underestimate you" or "your haircut makes your nose look smaller." Kind of.

Does *The Mandalorian* have an escape clause out of this use of parsec? Possibly. The conversation between The Client and Mando took place on the planet Nevarro. If the star system in which Nevarro was located was incredibly populous and practically swarming with bounty hunters for no apparent reason, The Client's comment might make more sense, but Nevarro was located in a sector of the Outer Rim Territories that was sparsely populated.

The best we can do in *The Mandalorian*'s defense is this: Following the fall of the Empire, the Bounty Hunters' Guild was situated on Nevarro. They were led by Greef Karga, who operated from a city cantina, which we see in "Chapter 1." By the time the New Republic was established, a murder of Mandalorians (assuming that Mandalorians have the same collective noun as crows) came to hang out in a covert in the city sewers (times were tough). With a planet full of sewer-dwelling bounty hunters, The Client was spoiled for choice, and "best in the parsec" wasn't as much of a backhanded compliment as it seems. Perhaps.

WHAT'S WITH ALL THE MONO-BIOME WORLDS IN *THE MANDALORIAN?*

One of the more commonly described oddities of the Star Wars Universe is the notion of entire planets with one climate. We are introduced to desert planets and ice planets and swamp planets. Of course, any one planet could have a myriad of different climates and ecosystems in the real world. But this peculiar detail reveals the metaphor that Star Wars is working with: each planet in *Star Wars* is basically each country in our world. This metaphor allows the Star Wars Universe to essentially retell stories from the era of British imperialism. The empire then becomes quite familiar to us, especially on the surface. In *Star Wars*, though, the empire is the enemy and the undisciplined, free-spirited rebels become the heroes—thus aligning *Star Wars* with thematic elements from the American Western even amid the trappings of British imperialist narratives.

—J. Andrew Deman, *The Conversation* (2017)

Star Wars Biomes

In 2021, and naturally on May 4th, Disney+ broadcast *Star Wars Biomes*. The viewer was invited to "take a virtual vacation to some of the *Star Wars* films' most iconic and beloved locations like Hoth, Tatooine, and Sorgan, as this charming series whisks you off for fly-through tours of the galaxy far, far away." This program is one of the more unique offerings on Disney+, with the dynamic tours feeling like a journey on a ship, and evoking fond personal memories of the indie video adventure game,

Journey, released for PlayStation 3 in 2012, in which the player controls a robed figure that floats around a vast desert.

Star Wars Biomes bears witness to several key moments in the *Star Wars* Skywalker saga and its three trilogies, as well as *The Mandalorian*, and showcases the galaxy's menu of curious worlds, some of which are explored in a new way in the program. *Biomes* is a one-off, presenting that *Journey*-like calming experience, and featuring just the barest hints of character or plot. We get mere glimpses of *Star Wars* worlds, which consist mainly of solitary biomes. *Star Wars* often gets flak for these mono-biome worlds, so let's take a deeper look and suggest that perhaps in part the flak is a little unfair.

Our journey begins on Hoth. A planet covered in ice and snow, glacial as frozen iron, with howling winds and icy drafts, a largely lifeless world with but a few species hardy enough to survive its constant snowball cold. Next is Tatooine. A planet of sand, of silence, with a pure clean atmosphere where, when the Sun is up, you can't tell where heaven ends and the ground begins. Third is a planet from *The Mandalorian*: Sorgan, like Endor and Corvus, also in *The Mandalorian*, is a world of forests and groves. A world of trees and farmers. Trees are sanctuaries, so it's little wonder Endor was known as the Sanctuary Moon. The fourth planet on the tour is Crait, a salt-covered world whose monotonous salt plains surface we see disturbed by Imperial walkers to reveal a red soil below, creating a most beautiful aesthetic. Fifth is fiery Mustafar, land of lava and Vader's sinister castle, which we can't possibly miss in our flyover of the planet's surface—an ominous world that witnessed Anakin's metamorphosis into Darth Vader. Finally, Ahch-To, a world that sounds like a sneeze, a planet wrapped in an ocean, as Earth once was, and a little like Trask in *The Mandalorian*, with sprinklings of scattered islands.

Empires in Space

Each of the worlds in *Star Wars Biomes* could be seen as symbolic of planet Earth, whether its future, present, or prehistoric past. Take Hoth, for example. Characteristic of Canada's status as coldest nation in the world, or emblematic of Snowball Earth, when half a billion years back, during one of our planet's icehouse climates, the surface of our world became entirely frozen? Tatooine. A symbol of Earth's Arabian deserts, or the shape of things to come if global warming gets its way? Sorgan, nostalgic of our lost and wooded home world of old, or dreams of a new Eden? Last, Ahch-To, evocative of the prehistoric time when Earth boasted one enormous ocean, or a future vision of a drowned Earth after sea waters inevitably rise?

There's a fascinating article on this topic in *The Conversation*, a network of not-for-profit media outlets publishing expert opinion and research reports online. Written by Professor J. Andrew Deman, the article suggests that the *Star Wars* franchise has always been far more interested in our past than in our future. Professor Deman claims that despite the spaceships and cosmic helmets, the kind of future *Star Wars* and *The Mandalorian* portray is far more reflective than speculative.

As Professor Deman points out, "We are taken away to the far reaches of space, but, once there, we find the Second World War, feudal Japan, 1930s America"—and, we might add, an obsession with the Old West in the case of the gunslinging bounty hunter, Din Djarin. As can be seen from Professor Deman's quote at the start of this chapter, the historical and reflective nature of the Star Wars Universe enables the franchise to retell tales from the days of the British empire.

Famous American science fiction author Ursula K. Le Guin had predicted such a state of affairs. Writing just two years before the first *Star Wars* film, Le Guin also suggested that the imperial metaphor was perplexing and typical of a tendency in science fiction to pine for the past rather than to aspire toward the future. She writes in her 1975 essay, "American Science Fiction and the Other":

> From a social point of view most science fiction has been incredibly regressive and unimaginative. All those Galactic Empires, taken straight from the British Empire of 1880. All those planets—with 80 trillion miles between them!—conceived of as warring nation-states, or as colonies to be exploited, or to be nudged by the benevolent Imperium of Earth toward self-development—the White Man's Burden all over again. The Rotary Club on Alpha Centauri, that's the size of it.

Dogfights and Dens of Bandits

Professor Deman's description goes one step further. Once established, he says, the imperial space of *Star Wars* and, we must admit, *The Mandalorian* is then populated with the furniture of the past rather than the tech of the future. Thus, the Universe's aerial combats are more akin to World War I dogfights than they are to modern aviation warfare with "unmanned drones and missiles that leave the aircraft long before the enemy fighter is even visible to the pilot." Saloons and cantinas for gunmen, gamblers, and outlaws of the Old West, rather than a prescient peek into the way we may have fun in the future. A Jedi culture by way of feudal Japan and an Empire with the reach of Britain of 1880, but bedecked with the imperial aesthetic of the Nazi Third Reich era.

Yet, whether by design or good fortune, there is some scientific credibility to the planetary worlds that populate space in *Star Wars*. Just think about the other planets in our Solar System. The so-called terrestrial planets, Mercury, Venus, and Mars, are all pretty much mono-biomes, in as much as there is anything "bio" about them. Mercury, a crater-laden world that has been geologically inactive for billions of years. Venus, a hot and sweltering toxic planet peppered with crunched mountains and myriad volcanoes, making Mustafar look like Eden. *The Mandalorian*'s Nevarro is a little like a watered-down, lamer version of Venus, albeit with a breathable atmosphere. Mars, a dusty and cold desert world, but also dynamic with seasons, polar caps, and canyons, with evidence it was warm and wet in the ancient past. (Give or take the odd climatic quirk, Arvala-7 in *The Mandalorian*, among others, is quite like a breathable variant of Mars.)

Just Like Bespin, Endor, and Yavin

Then there's our gas giants, Jupiter and Saturn. Jupiter, with its stripes and swirls, those cold and windy clouds of ammonia and water, floating in an atmosphere of hydrogen and helium. The mighty planet's iconic Great Red Spot, a huge storm big enough to swallow the Earth with ease, a hurricane that's raged for hundreds of years. Saturn, blanketed by clouds whose clues appear as jet streams, subtle stripes, and storms. A world of many shades of yellow, brown, and gray, Saturn's atmospheric winds reach 1,600 feet per second in the equatorial regions. And those rings. Ranging in size from tiny, dust-sized ice grains to hunks as big as a house, Saturn's rings are a spectacular sight. Smithereens of a once-orbiting Moon, a few of the masses are as large as mountains. Seen from the cloud tops of Saturn, the rings would bling mostly white. And each ring orbits its world at a different speed.

Jupiter and Saturn are both orbited by dozens of Moons. At the time of writing, the latest research confirms that Saturn has eighty-three satellites with formal designations, and Jupiter ninety-one. Each gas giant has innumerable additional Moonlets. It's easy to see why each planetary structure is considered a Solar System in miniature. Take the Moons of Saturn. Some of its more substantial Moons are worlds in their own right. For example, half as large again as our own Moon, and larger than the planet Mercury, Saturn's Titan is the second-largest Moon in the Solar System. Boasting one of the most Earth-like places in the solar neighborhood, albeit at hugely colder temperatures and with a contrasting chemistry, Titan is so cold (-290°F) that water ice plays the role of rock.

Fly Me to the Moons

Jupiter's main Moons are also worlds in their own right. The nearest world to Jupiter is Io. Io has a climate like Nevarro and Mustafar, but far from breathable! With a surface littered with hundreds of volcanoes, many spewing forth sulfurous plumes hundreds of miles high, Io is the most volcanically active world in the Solar System, with lakes of molten silicate lava on its surface. What are the conditions that provoke this mono-climate? As Jupiter's third largest and the innermost Galilean satellite, Io is caught in a gravitational trap, a tug-of-war between Jupiter and two nearby Jovian Moons, Europa and Ganymede, the largest Moon in the Solar System.

Europa, too, is pretty mono-climatic. Like Earth, Europa is believed to have an iron core, a rocky mantle, and an ocean of salty water. But that's where the similarity ends. Unlike Earth, Europa's ocean lies deep below a shell of ice, probably ten to fifteen miles thick, with an estimated depth of forty to one hundred miles. (Imagine an ocean one hundred miles deep; the Mariana Trench, the deepest part of Earth's oceans, is only seven miles deep.) While evidence for Europa's internal ocean is strong, confirmation awaits a future mission.

Meanwhile, a little like Crait, Europa's surface is crisscrossed by long, linear fractures. But whereas Crait's surface is salt-covered, Europa's is ice. Along Europa's countless fractures, and in blotchy patterns across its surface, is a reddish material whose composition is unknown, but likely consists of salt and sulfur compounds that have been mixed with the water ice and modified by radiation. This surface composition may hold hints to Europa's potential as a habitable world.

Voyager

Here's the thing about all these mono-climatic worlds in our Solar System. How much did we know about them when the first *Star Wars* movie was made? Not a lot. The unveiling of the Solar System truly took off with Voyager, the American scientific program that sent out two robotic interstellar probes, *Voyager 1* and *Voyager 2*, that weren't launched until 1977 itself. Their joint mission was to take advantage of a favorable alignment of Jupiter and Saturn, to perform fly-bys and send data back to NASA. Indeed, the Voyager probes are out there still, beyond the boundary of the Solar System and deep into interstellar space. As US author Stephen J. Pyne wrote, "Voyager did things no one predicted, found scenes no one expected, and promises to outlive its inventors. Like a great painting or an abiding institution, it has acquired an existence of its own, a destiny beyond the grasp of its handlers."

It wasn't until March 1979, the same month that principal photography began on *The Empire Strikes Back*, that *Voyager 1* closed in on Jupiter. Pulled by gravity, the probe hit 80,000 mph. The images it sent back to Earth were simply astounding. The surprises started with Jupiter's Moons. Io had volcanoes, the first found beyond Earth. Europa was covered in cracks. Callisto was heavily cratered and crusted and seemed lit from within. This giant planet, a marble world spinning in a black void, with its great red storm swirling and raging, made front page news across our globe.

Those Galilean Moons that Galileo first spied as mere blurs through his telescope way back in 1610 had finally become real planetary places. Performing to perfection, Voyager sent back roughly twenty thousand close-ups before using Jupiter's gravitational impetus and winging its way to Saturn. And in November 1980, with *The Empire Strikes Back* a mere six months old, Voyager began beaming back photos of the Saturnian system. Gargantuan rings of cosmic bling. Dazzling Moons. And awe-inspiring views of Saturn, merely a mono-biome ball of swirling gas, yet breathtakingly beautiful.

The Voyager program was the gift that kept on giving. Why? Because, expected to last only a decade, Voyager just kept going. When *Voyager 1* reached the limits of the Solar System, American astronomer Carl Sagan urged the mission team to capture the poignant moment, so the techs turned around the spacecraft cameras to take a picture of the whole Solar System. And there among the 640,000 pixels was the single Pale Blue Dot of Earth. Sagan said, to a stunned world:

> Look again at that dot. That's here. That's home. That's us. On it everyone you love, everyone you know, everyone you ever heard of, every human being who ever was, lived out their lives. This distant image of our tiny world underscores our responsibility to deal more kindly with one another, and to preserve and cherish the Pale Blue Dot, the only home we've ever known.

Earth as Analog

Here's the thing about those mono-climates on *The Mandalorian* and in the *Star Wars* Universe. Sure, if you use planet Earth as your analog, as your comparison, fictional alien worlds are not going to compare well. But as we have seen through the space exploration of the Solar System and, by extrapolation, exoplanets in deep space, other Galaxies have their fair share of mono-biomes, even if they are not all breathable to humans. Maybe we should give *The Mandalorian* a break. The creators of the Star

Wars Universe took an intelligent punt on planetary worlds back in 1977, a time before we had discovered very much at all about the planetary worlds and Moons in our own backyard.

Finally, on the importance of the Earth analog, consider this. Back in 2009, the American space observatory Kepler was launched. Its mission was to find Earth analogs, Earth-like planets orbiting stars other than our Sun. Kepler took off four hundred years after Galileo first used the telescope to discover Jupiter's Moons, and is of course named after the first great Copernican theorist, Johannes Kepler. At the time, and based on Kepler's early findings, Seth Shostak, senior astronomer at the SETI institute, estimated that "within a thousand light-years of Earth," there are "at least thirty thousand habitable planets." And based on the same findings, the Kepler team projected that there are "at least 50 billion planets in the Milky Way," of which "at least 500 million" are in the habitable zone. NASA's Jet Propulsion Laboratory was of a similar opinion. JPL reported an expectation of two billion "Earth analogs" in our Galaxy and noted around "50 billion other galaxies" potentially bearing around one sextillion Earth analog planets. Maybe some day in the future we will discover an Earth analog, that holy grail of planet hunting.

WHICH MANDALORIAN ALIENS SHOULD NEVER HAVE MADE THE SCIENCE CUT?

The Science of Aliens touring exhibition that launched at the London Science Museum in October 2005 asked the question, "Are we alone in the Universe?" The exhibition looked at science fiction archetypes to tell us about the real possibilities for alien life. A variety of experts gave advice on the exhibition development including Dr. Jack Cohen, Professor Simon Conway Morris, and myself.

The Assumption of Mediocrity

Earth is an ordinary planet moving around a local star we call the Sun. This statement doesn't seem revolutionary, and certainly not heretical, yet at one time in the past it most certainly was both. After Copernicus published his famous book in 1543, a so-called Copernican Principle was established. This Principle can be summed up with the simple phrase: if Earth is not central, then neither is humanity.

More fully, we know that for many centuries the assumed prime position of cosmic centrality assigned to the Earth was a combination of bad astronomy and theology: our world is as important as we are, made in God's image, and we wield dominion over the whole planet, creatures, continents, and seas. Thus, the Copernican displacement of the central Earth was an act of mutiny, crucial to the Scientific Revolution and to the Enlightenment.

Another aspect of the Copernican Principle was the assumption of mediocrity. This is related to the question of life in the Universe, and it goes something like this: if our planet is not unique, neither is life on Earth.

Given that there *is* life on planet Earth and planets have been found to be common, life should be abundant at least on terrestrial planets in the Universe, including intelligent life. In short, the assumption of mediocrity suggests that life is so abundant in deep space that it's a mediocre property of the Universe. And the *Star Wars* franchise, of course, profoundly recognizes this assumption.

The Science of Aliens

In my *Science of Aliens* book, I wrote about the way in which writers and moviemakers try to imagine the unimaginable when it comes to alien life. Why is the task so difficult? Simply because nowhere on any alien world in deep space would our eyes rest upon the familiar: we would not recognize any plants or trees or any of the animals from our own world. Alien life would be truly strange, stranger than the nightmarish creatures of Earth's deepest oceans, stranger even than the terrestrial insect empire, the details of whose horrors were historically hidden from humans by their sheer infinitesimal size.

It's pretty tricky, imagining the unknown, but the *Star Wars* franchise has come up with some pretty cool ways of imagining the unimaginable. Their creative minds have populated silver screens large and small with a veritable zoo of extraterrestrials, from the dread-inducing sando aqua monster ("there's always a bigger fish!") to the drop-dead delight of Baby Yoda.

Given all this effort, it's hardly surprising that *Star Wars* writers and moviemakers sometimes fall short of such high standards. Now, don't get me wrong. We aren't talking historically badly rendered aliens here. Nothing as ridiculous as *Men in Black*'s race of "unique" aliens known as the Ballchinians, whose long sagging faces (at least on the males) are due to the genitalia on their chins. Nowhere near as dumb as the extraterrestrials in the 2002 M. Night Shyamalan movie, *Signs*. These aliens, who decide to invade a planet which is mostly water, and eat humans, who are mostly water, have a main weakness. Guess what that weakness is? Yup, that's right: water. They came halfway across space with such little research? Didn't they notice the ocean details on atmospheric entry? And, for that matter, there's copious water in Earth's atmosphere, so these dumbest of aliens would have started feeling the burn before they landed.

Where did *The Mandalorian* get it wrong? Which of this program's aliens should never have made the science cut? Let's take a look at the season three opener of *The Mandalorian*, "Chapter 17: The Apostate."

Pirates

The first problem that pops up is the pirates. Mando and Grogu have headed to Nevarro, where they have reunited with the now High Magistrate Greef Karga. The prospect of Mando and Karga working together is interrupted by a band of pirates who were previously members of Karga's guild. So far, so good.

But what's this? The pirates look like various humanoid versions of terrestrial sea creatures. Their ringleader, Vane, bears an uncanny resemblance to a kind of juvenile version of Davy Jones from the Disney movie production, *Pirates of the Caribbean: Dead Man's Chest*. Unlike Davy Jones, Vane's tentacles are not yet fully matured, but the similarity is still striking. And, just in case we miss this fact, standing behind Vane *is* a pirate with a full set of mature octopus-like tentacles.

If this isn't enough for the already patronized viewer, later in the episode we get Captain Gorian Shard, leader of the gang of pirates in the Nevarro sector. Shard is also pedantically dressed like a terrestrial pirate. He's not so much a swamp thing as a sea monster. He looks like his head has been dredged across the ocean floor. He is dreadlocked in seaweed. He looks like an option on a Chinese menu alongside the crispy duck. The writers couldn't have come up with something a little more imaginative than aliens who look like terrestrial sea creatures dressed like the cast of *Pirates of the Caribbean*? Perhaps there is some universal law of piracy, that all people who attack and rob ships must defer to the preferred fashions of *sea pirates*, whether or not the ships they raid sail on oceans or in deep space.

The most offensive thing about Shard, to the educated viewer, is the idea of Shard being a "Pirate King." On Earth, during the so-called Golden Age of Piracy that ran from the late seventeenth century to the early eighteenth, sea pirates were communities that operated outside of the established legal systems of the time. Pirate communities were famous for their self-governance and nonhierarchical structures. They organized democratically, with pirates electing their own leaders and making decisions collectively. Pirate crews were mostly made up of sailors and other individuals who had been mistreated or oppressed by the ruling classes of their time. Pirate crews often operated under a code of conduct that was agreed upon by all crew members. This code, often known as the "articles of agreement," normally included arrangements for the fair treatment of all crew members, democratic decision-making, and a system for dividing booty and plunder evenly among the crew.

Pirates deliberately limited the captain's power and kept order on board the ship. A pirate captain was elected by *all* the crew and could be replaced by a majority vote by the same method. Cowardly or brutal captains were swiftly voted out. And all this was a way to reduce conflict among the pirates and maximize their profits. We can easily imagine that the nature of piracy would dictate similar arrangements on other worlds, with the notion of kings sent to oblivion. Sadly, we get "Pirate King" Shard instead.

The Anzellans

Gladly putting the inappropriate pirates to one side, next up in "Chapter 17: The Apostate" we meet an alien race known as the Anzellans. Mando declares that he intends to revive IG-11. He needs a droid he can trust by his side on Mandalore. IG-11's parts are assembled from his memorial statue, and IG-11 is revived. The trouble is, IG-11 reverts to his initial programming and tries to kill Grogu. After shutting IG-11 down, the help of Anzellan mechanics is enlisted.

The Anzellans, we are elsewhere informed, were a diminutive sentient species so tiny that they often took up tech jobs. We can only marvel at the sophisticated thinking here. Intricate and nano-electronic wizardry clearly needs nano-aliens. Little guys for little jobs. Why not?

Moreover, if this reason wasn't enough (and it clearly isn't!), we are provided with the conveniently bolted-on afterthought that the eyes of Anzellans had floating corneal micro lenses that could see microscopic details. We can only wonder at what evolutionary reason led the Anzellans to such microscopic sight. Then again, rather than taxing ourselves about the history of Anzellan natural selection, we could just remind ourselves that so-called super senses are something of a cliche in that silliest part of sci-fi: comic books. Super creatures are forever blessed with senses keener than those of a mere human. The classic five manifestations are smell, hearing, taste, touch, but most of all, that sovereign of the senses, sight. Thus, super sight comes in many different flavors: night vision, infrared vision, slo-mo "bullet time," and, yes you guessed it, telescopic and microscopic vision.

One of the reveals about the Anzellans in "Chapter 17: The Apostate" is that they are the best droidsmiths in the Outer Rim. Once more, we can only wonder what makes the Anzellans better droidsmiths than actual droids created and engineered to do that exact task. After all, we know there are repair droids, also known as mechanic droids. Surely a society and a culture progressive enough to develop sophisticated droids can muster maintenance droids who can repair their own kind a little better than diminutive bug-eyed critters.

Purrgil

Leaving the Anzellans to their microscopic mechanics, next up in "Chapter 17: The Apostate," the franchise's confusion between deep sea and deep space reaches fever pitch with those most enigmatic of beasts, the Purrgil. The story goes like this. Mando and Grogu are heading for Nevarro. On part of the journey through a hyperspace tunnel, Grogu catches a tantalizing glimpse of several dark shadowy shapes in the blur of spacetime. Curiously, the huge shadows are whale-like in form, and belong to mysterious creatures known as Purrgil. And it's very likely that the Purrgil shoot right to the top of our daft aliens list.

The Purrgil, we are told, were a semi-sentient species of mighty whales that lived in deep space, journeying between star systems. Ordinary Purrgils were around the size of a small starship. But Purrgil Ultra, a subspecies, was appreciably bigger. All species of Purrgil had streamlined bodies, smooth purple skin, four large hind tentacles, and numerous fins. The head of a Purrgil had a typically bulbous form, with an eye on each side. Traveling in groups comprising dozens of individuals, Purrgils were able to fly through hyperspace by creating simu-tunnels. Indeed, we are also told, it was the Purrgils' natural ability to breach hyperspace that inspired sentients to develop hyperdrive tech, and also led to sentients creating way-finder tech for hyperspace navigation.

We know little of the evolution of the Purrgil, other than the fact that they lived in groups called pods or flocks. These groups of dozens of individual Purrgils were led by a bigger specimen known as the Purrgil King. (Not content with misapplying feudal power structures on the pirates, we now have kings among the whales. For comparison, the social structures of Earth's killer whales are way more complex, with most organized in matriarchal societies.) As a semi-sentient species, Purrgils exhibited a modicum of intelligence, though they could also rather carelessly inflict unintentional harm by crashing into starships.

Okay, so we all know whales are cool. That's a given. And sure, it's probably the case that space whales would be even cooler. In *Dune*, author Frank Herbert talked about the House Harkonnnen "manipulating space whale fur prices." *Doctor Who* gave us a star whale, a creature that came to help humanity, but was turned into a slave for its troubles. The humpback whales of *Star Trek IV: The Voyage Home* actually turn out to be aliens from another world. And, as if to pour scorn on the very idea of all these space whales, in his *Hitchhiker's Guide to the Galaxy*, Douglas Adams has a whale called into being with little time to contemplate its perilous situation in midair before going "splat."

Foxes and Whales

One hates to be pedantic about such enigmatic creatures as the space whale, but the truly enquiring mind can't help wonder how on Earth such an oceangoing species ended up in space. Let's take the example of Iceland. The landmass we now know as Iceland began to form around sixty million years ago when tectonic plates began to pull apart and enough lava piled up to make land. But how did Iceland get its flora and fauna?

We know that Iceland had but one native land animal when the Norse first arrived between 870 and 930 AD. The arctic fox is believed to have arrived in Iceland at the end of the last Ice Age by simply trotting across the frozen ocean from Greenland and Scandinavia. We have the UK's University of Durham to thank for this research. Their scholars found that, during the Little Ice Age's dip in temperatures between two hundred to five hundred years ago, a new wave of arctic foxes colonized Iceland by using a "bridge" of sea ice, allowing them to migrate from different Arctic regions in Canada, Russia, and Greenland. And if the arctic fox can migrate in the brevity of the Little Ice Age, then it can certainly do so during the millennia of the Ice Age itself. Today, of course, we find multiple species across Iceland. But none other than the arctic fox arrived naturally. The rest were either deliberately brought over by humans or somehow snuck across on boats.

The Secret of the Space Whales

What kind of evolutionary mechanism can we imagine for whales migrating into space? Who knows, maybe *Star Trek* has a point and the whales have been aliens all along, conducting a sophisticated series of experiments on humans (like the white mice in the *Hitchhiker's Guide to the Galaxy*). But all joking and ridicule aside, there *is* a possible scenario by which the whales could have ended up in space.

Consider the following science fiction movie classic: Arthur C. Clarke and Stanley Kubrick's *2001: A Space Odyssey* (1968), described by Kubrick as a "scientific definition of God." It is a story of the effective creation and resurrection of humanity under the episodic guiding hand of superior alien beings. Kubrick's motion picture traces humanity's journey through three stages. The journey begins with Stone Age humans. A small band of human-apes are on the long, pathetic road to species extinction. But they are visited by mysterious and elusive aliens: an artifact in the shape of a black monolith. The mysterious presence of the monolith transforms the hominid horizon. Ape becomes human, and humanity's ultimate journey to superman begins.

When humans evolve to the space age, the potent evolutionary force of the alien monolith is triggered once more. Finally, the odyssey of self-discovery culminates under the watchful presence of the alien monoliths when modern humanity, in the form of an individual astronaut, comes to an end. The movie screen replete with the massive presence of planet Earth, the fetus of a superhuman star-child floats into view. The star-child moves through spacetime without artifice, the image suggesting a new superhuman power. Humanity has transcended all earthly limitations.

You see my point? Sure, it's a solution based more in science fiction than fact, but if Kubrick can imagine an alien intelligence so immense that it can bestow superhuman powers on humanity, why can't super-whale status be bestowed on the purrgils? (Notwithstanding the scientific unlikelihood of whale-like creatures such as the purrgils evolving on a planet other than Earth, of course.)

Picture the scene. A race of purrgils are minding their own business in the depths of their home ocean, foraging and exploring, migrating and mating, socializing and sleeping, and singing songs to their children. Up pops the same kind of interfering-but-elusive aliens as featured in *2001: A Space Odyssey*, which eventually leads to the purrgils, like humans in Kubrick's movie, being able to move through spacetime without artifice. And even hyperspace. We just don't know. It might happen!

WHAT IS THE FRONTIER SPACE OF MANDALORIAN ASTROGRAPHY?

Astrography (noun): The art of describing or delineating the stars; a description or mapping of the heavens.

> The *Star Wars* galaxy is . . . divided into regions, with the Deep
> Core as the central and most luminous region of its space. The
> populous Core Worlds are where the human species first evolved.
> The Core includes planets such as Alderaan and Coruscant,
> which were granted permission to settle new planets. They were
> followed by the Colonies (which included the planets Castell and
> Halcyon), the Inner Rim (which housed the world of Onderon),
> the Expansion Region (housing the likes of Aquaris), and the Mid
> Rim (which included Naboo) and Outer Rim Territories (which
> included Hoth and Tatooine).
> —Mark Brake and Jon Chase, *The Science of Star Wars* (2016)

Galaxy and "Galaxy"

Is there life elsewhere in our Galaxy? No one yet knows. But the evidence suggests there may well be. In my coauthored book, *The Science of Star Wars*, I talk about how *Star Wars* could help us resolve this question about life in our Galaxy with the franchise's huge uncharted area called the "Unknown Regions." This area, which lay to the galactic west, remained mostly unexplored throughout history, as most trade routes headed outward in the direction of the galactic east. Not only that, and this is of particular interest to us regarding *The Mandalorian*, but there was also a region at the very edge of the *Star Wars* galaxy known as Wild Space, a region inhabited by sentient species but never fully charted, explored, or "civilized."

This *Star Wars* scenario leads us to an interesting conclusion about the Milky Way. We haven't yet seen any alien life because Earth might reside in our Galaxy's own version of the Unknown Region. Our Solar System is sitting way out from the core of the Milky Way, which, for all we know, may have its own version of the fashionable Core Worlds. It would take a journey of trillions of miles, either through the dense Galactic Core, or through its spiraled suburbs, to get to planet Earth.

This *Star Wars* scenario also provides us with an adequate solution to the Fermi Paradox, neatly summed up in the following sentence: if alien life exists in our Galaxy, "where *is* everybody?!" In other words, and more fully, if it's safe to assume that we humans have a lot of cosmic company, and there really are a lot of alien civilizations out there in deep space, then surely some of them would have spread out, as is the case in *Star Wars* galactic history. Assuming such alien civilizations have had enough time to populate the Galaxy with their presence, why aren't their legions of sleek spaceships tearing across our skies? Where are the TIE fighters, X-Wings, and Imperial Class Star Destroyers? Again, the answer might be: Earth sits in the Milky Way's Unknown Region and the Empire hasn't yet come calling.

Galactic Math

There's another aspect to this question of Earth being in the Galaxy's Unknown Region. And it's this. Even if the Empire *had* come calling, or at least some kind of alien civilization explored our previously unknown regions, it might still mean our earthly civilization would not be discovered. How might that work? Scholars have built migration models for the expansion of alien civilizations such as the fact that alien planetary civilizations may only endure for around a million years, or that maybe only 3 percent of the Galaxy's star systems are actually habitable.

When scholars use such values, their models predict that a habitable world like Earth has a 10 percent chance of *not* having been visited in the past million years, so it's hardly a surprise that we should find ourselves isolated and unvisited. Moreover, these models also suggest that, like the *Star Wars* galaxy, there are other regions where alien visitors and extraterrestrial neighbors are perfectly possible, so scholars don't need to render the model extreme to produce a plausible scenario for alien life to be commonplace. All that's needed is an eye for detail on habitable planet numbers and the nature of stellar dynamics in the Galaxy. Instead of looking for individual exoplanets, astronomers should search for those stellar populations, those regions of the Galaxy where the architecture of space might promote interstellar planet hopping. So, the lesson to be learned both from *Star Wars* and the academic models is the same: it's totally typical for a planet like Earth to have remained unvisited, as yet, by alien civilizations. Planet Earth sits isolated in space, a world in waiting.

Wild Space and the Wild West

How much of the above discussion can help us with the astrography of *The Mandalorian*? Well, we know that the planet Mandalore had its own regional space empire in ancient times, led by their formidable warrior culture. So-called Mandalorian space is technically located in the galactic northeast of the Outer Rim, yet Mandalorian space is also reasonably close to the border with the Mid-Rim, which holds generally industrialized though not particularly important planets such as Naboo and Kashyyyk that are more developed than the true "frontier" worlds of the Outer Rim.

According to *The Mandalorian*, Mandalorian space was ultimately devastated by the Empire, with surviving enclaves scattering across the Outer Rim. *The Mandalorian* also produced the first franchise on-screen, live-action dialogue to confirm the official galactic astrography of the *Star Wars* canon. In "Chapter 12: The Siege," the second season's fourth episode, a school classroom is briefly shown on a planet in the Outer Rim. A lesson on galactic astrography is in process, tutored by a protocol droid. The droid confirms the order of the galaxy's major regions from the Deep Core to Wild Space: In the words of the droid, "Who can name one of the five major trade routes in the galaxy? The Hydian Way [runs] from the Outer Rim to as far away as the Core Worlds. However, there are several other regions within our galaxy. They are the Mid Rim, the Expansion Region, the Inner Rim, the Colonies, the Core, and the Deep Core."

The Outer Rim as Frontier

How does Mandalorian space in the Outer Rim compare with typical tropes of the American Old West? The so-called "Wild West" was the region west of the Mississippi River, roughly during a brief period in the second half of the nineteenth century. Some scholars take the start of the Wild West era as being the California Gold Rush of 1848, concluding in 1890 with the Census Bureau's official recognition of the end of the frontier. This period also includes the years of the American Civil War, which ran between 1861 and 1865. The geographical space west of the Mississippi is the setting for "the western," a stylized fictionalization of the American Old West.

Samuel Fuller, director and screenwriter of numerous Western films, had an interesting thing to say about them:

> I love the West. I read a lot about the West, and I'm shocked, I'm ashamed that in pictures they have not made the true story of the winning of the West—comprising 90 percent foreigners, 100 percent laborers, nothing to do with guns. Streets, mountains, roads, bridges, streams, forests—that's the winning of the West to me. Hard! Tremendous, tremendous fight. But [instead] we have, as you know, cowboys and Indians and all that.

Fuller is addressing the fact that the *real* Old West was nothing like the fictionalized cowboy version of nineteenth-century dime novels and twentieth-century movies. There were few mass shootouts. Quick-draw duels were rare. And violent gun-toting lawbreakers were not restricted to the deserts of the West. Moreover, as the guns of the day lacked accuracy, shootouts occurred at considerably closer proximity than they do in films. A quite shocking statistic is that the average western town had only one and a half murders each year (I'll avoid the obvious joke that the half must have been a cowboy relatively short in stature) and most of those murders weren't done with guns. In fact, actually *carrying* a gun in western towns was far more likely to get you arrested than shot. Your average citizen of the West was more likely to die from diseases like dysentery, cholera, and tuberculosis, or from being dragged through the pitiless dirt by their own horse, than to be murdered in the mayhem of a raging gunfight, or by a fatal scalping by Indians. In short, the Wild West was not so much wild as it was boring and dull.

It's the over-fictionalized cowboy version of the frontier West that we get in the *Star Wars* galactic geography. This is underlined to us from the get-go when Luke Skywalker complains of Tatooine with poetic exaggeration: "If there's a bright center to the Universe, you're on the planet that it's farthest from." Nonetheless, the worlds of the Outer Rim were distant from the galactic Core and Coruscant. Rival powers like the Sith and the Mandalorians contested this sector of space like the cowboys and Indians of the old Westerns. And just as civilized government must have felt very distant during the days of the Old West, to some extent the Republic kept out of Outer Rim affairs, even allowing slavery to flourish on planets like Tatooine.

Just as the "big four" Californian railroad tycoons decided to build the Central Pacific Railroad, with a concomitant unethical accumulation of geographical space, the Outer Rim had its criminal cartels, such as the Hutts and Black Sun, who often achieved dominance over whole planetary- and star-systems of galactic space. Government and Empire allowed both to continue without intervention. Moreover, the relative lawlessness of the Outer Rim was typified both by the facts that the Death Star was constructed here, far from the attention of the Galactic Senate, and numerous battles of the Galactic Civil War were fought in this sector of space.

Why Is the Outer Rim a Wilderness?

The Mandalorian has emphasized just how much the Outer Rim was a wilderness compared to much of the rest of the galaxy. Though, just like the actual Old West, there were pockets of civilization within that wilderness. The distal location of the Outer Rim was naturally one reason why this region was regarded as a wilderness. It's well known that an empire's authority atrophies over distance, so distant planets became natural homes to crimes that often went unchecked, and the galactic government of the day was less likely to commit resources to bring errant planets under control. After all, the New Republic's focus remained on the massive centers of civilization of the Core Worlds, and that focus blurred by the time you journeyed far in the galaxy away from those key planets of Republic or Empire.

How did the human populations of the Outer Rim wilderness develop in the first place? Here, another analogy with the Old West is possible. Before Columbus, the continent of the Old West was already populated. The indigenous peoples hadn't always been there, of course, nor had they originated there. But most scholars believe the indigenous peoples had occupied "American" lands for around twenty thousand years. Much later, from the fifteenth century onward, Europeans began arriving on the same landmass with their own culture and agendas, and they met with other cultures and races.

There is a parallel story in *Star Wars*. Humans spread across the *Star Wars* galaxy using the hyperdrive and their scouts and explorers journeyed and met with other species in a time called the Expansionist Era. Human colonies, which had migrated through the Core Worlds through slower-than-light travel, later founded daughter colonies of their own in what became the Colonies region. When faster-than-light travel evolved, those regions closest to the galactic Core were colonized first, allowing them more time to evolve and develop compared to regions more removed from the center of the galaxy.

There is much western influence in the portrayal of Mandalorian space. As the largest charted region in the galaxy, it was replete with diverse worlds and rough and ready, primitive frontier planets. Only Wild Space and the Unknown Regions were wilder. And given that we may well live in our Galaxy's own version of the Unknown Regions, maybe that should give us pause for thought!

PART III
TECH AND TIME

DO WE HAVE THE TECH TO BUILD MANDO'S HELMET?

"I'm a Mandalorian. Weapons are part of my religion."
—"Chapter 2: The Child," Season 1 of *The Mandalorian* (2019)

"No living thing has seen me without my helmet since I swore the creed."
—"Chapter 8: Redemption," Season 1 of *The Mandalorian* (2019)

"All this talk of the Empire, and they lasted less than thirty years. Mandalorians have existed ten thousand."
—"Chapter 5: Return of the Mandalorian," Season 1 of
The Book of Boba Fett (2021)

Helmet Happy

Where would the mythology of *Star Wars* be without helmets? If this was ever in doubt, consider the recent intro for Disney's *Star Wars* properties. The intro sequences for titles such as *The Mandalorian*, *Obi-Wan Kenobi*, and *Andor* on Disney+ are visually summed up in a series of tech images. First, we get a chrome-gleaming Lucasfilm Ltd. logo. Then, we get a sequence of laser-lit stills of the iconic helmets of *Star Wars* mythology, including Darth Vader, R2D2, C3PO, a stormtrooper, and a Mandalorian helmet.

The infamously helmeted Darth Vader is one of cinema's greatest ever villains. In June 2003, as part of the American Film Institute (AFI) 100 Years series, they compiled a list of the screen's top fifty villains.

Darth Vader came in as third greatest villain, behind human psychos Dr. Hannibal Lecter and Norman Bates. But for *Star Wars* fans, Darth is eviler still; all the more terrifying and unknowable due to the anonymity and mystique of his faceless and near-featureless helmet.

Flush with the success of old Darth, *Star Wars* then created that other helmeted villain, Kylo Ren. As the master of the Knights of Ren, Kylo's custom and combat-grade headgear strikes a new chord of fear and dread in the galaxy. Rather than the glossy finish sported by Darth, the Kylo headgear has a weather-beaten look, which speaks of the horrors its wearer has unleashed on his way to power. But *Star Wars* doesn't stop at Darth and Kylo. Other iconic helmet wearers include the original trilogy's Imperial stormtroopers, with their sub-Nazi look; the clone troopers, with their top-finned helmets, complete with Mandalorian-inspired visor; and the death troopers, with their bone-chilling, don't-fuck-with-the-Empire, black-on-black look.

That Legendary Look

For many *Star Wars* fans, the coolest headgear belongs to the Mandalorian culture. That incredible armored design. That T-shaped visor. That drop-down antenna. That associated legend of Boba Fett. That matchless beskar build. A spec meant for clan members and bounty hunters alike, the mythical helmet from Mandalore was the sickest headgear a militia could muster. To take the design just one step further, Mandalorian super commandos even sometimes adorn their helmets with horns.

Mythical is the word here. The look of the Mandalorian helmet owes much to our earthly past. On our planet, helmets are as old as war itself. Headgear was among the very first forms of protection that humans built for battle. Turn time back a full forty-three centuries to the ancient days of Akkadians and Sumerians, and you'll find that their warriors wore helmets. As weapons evolved with warfare, so did the helmets used in combat. Headgear fashioned from leather evolved into brass, bronze, and steel.

The helmet design of the Mandalorian culture bows down to ancient Greece's iconic Corinthian helmet. Named after the city-state of Corinth, this helmet is thought to be the most popular helmet of the archaic and early classical period. It was the helmet of choice for the Greek hoplites, for example, the citizen-soldiers of ancient Greek city-states. As ordinary Greek citizens, the hoplites lacked the military training of professional soldiers, such as those later in the Roman army, but made up for it with their war formations, using the phalanx formation to be effective in war with fewer soldiers. The hoplites appear in some of history's most famous ancient battles such as Marathon, Thermopylae, and Plataea.

Built in bronze, and particularly popular between the sixth and first century BC, the Corinthian helmet covered almost the entire head, commonly only leaving gaps for the eyes, nose, and mouth, forming that iconic T-shaped aperture we also see in *The Mandalorian*. The design of the helmet meant that vision and hearing were restricted. While this would have been possibly fatal for a normal soldier, as hoplites operated in that tight phalanx formation, their comrades on either side did the seeing and hearing for them. The helmet also boasted a large, curved projection that protected the nape of the neck, a potentially vulnerable attack point in battle.

Bronze and Beskar

We know from writings of the ancient Greek author and poet Homer that the helmet was critical to the survival of the hoplites in combat. The vast majority of Corinthian helmets were fashioned from a single sheet of bronze. And for good reason. This technique created a seamless helmet.

In Mandalorian culture, the station of the Armorer was a prestigious one. The ancient Greek world also held their blacksmiths in high regard. After all, to build a seamless Corinthian helmet from a single sheet of bronze deserves respect for the smiths of the era. Indeed, Greek blacksmiths, metalworkers, sculptors, and metallurgists had their own god in Hephaestus.

Just in case you thought the ancient world wasn't all joined up, like the world is today, consider this example of the importance of metals and metalworkers in the ancient Greek world. The Minoan civilization, to be precise, which flourished between about 3650 BC and 1400 BC. Homer said there were as many as ninety cities on ancient Minoan Crete alone. Minoan civilization shows how metal riches began to circulate through trade. The Minoans' merchant fleet sailed for hundreds of miles, from Syria in the east to Britain in the west, where some scholars believe they may have gotten their tin. At the time, Britain was known as "the tin islands." That's a journey of thousands of miles, just in search of the tin that sits in the Corinthian helmet, though Corinth began to develop as a commercial center much later.

While bronze is no competition for beskar, the Corinthian helmet was certainly capable of *deflecting* arrows, even if it couldn't stop a direct arrow shot. (It makes you realize just how fast those ancient arrows must have flown to be able to pierce armor.) The bronze build also had the benefit of providing protection from edged weapons. Archeologically preserved Corinthian helmets weigh between two and four and a half pounds. That's similar to modern military headgear. The ancient Corinthian helmet tended to be custom made, and was often lined with leather, felt, or sponge for comfort. It was also common for the helmet to be worn raised on the crown of the head, tipped casually upward while not in combat, to increase comfortability and give the wearer that jaunty, relaxed look. (This fashion was so well established that it became a decorative feature of Italian Corinthian helmets later on.) One probable reason for tipping the Corinthian was its bronze makeup. Bronze has a lower conductivity than other metals, but it still conducts and loses heat, so the hoplite warrior would grow warm or cold with the weather and/or heat of battle.

Interlude: The Cinematic Helmet

Before we compare Mando's helmet with its ancient Greek counterpart, let's look at how clever *The Mandalorian* has been in its use of this long-standing staple of screen costuming. For many decades, cinema has had an issue with the helmet. Robust headgear may protect our hero from harm, but it also hides his face from the audience. As the hero is the focus of emotion, we normally think that we need to see the actor's face to be able to feel our way through a scene.

Movies typically get around this problem by either making the character lose his helmet, or writing into the scene some kind of reason why the helmet must come off. Our hero whips off his helmet so we can center in on his face and read the emotion and tension of a scene, which would have been a pretty tricky ask, had his face been covered in iron. After all, this isn't so hard to believe. In real life, we can readily imagine that our hero would wish to convey his emotions to his comrades by taking off his helmet. But is there another creative way to deal with this helmet issue?

The Mandalorian has found one. For nearly every scene of the show, our hero, Din Djarin, wears a helmet. The Mythrol gives us advanced warning that we won't be seeing Mando's face anytime soon when he asks, in the very first episode, "Is it true that you guys never take off your helmets?" But the amazing thing is that we still empathize with Mando. We still feel the drama and tension of the tale, despite hardly ever seeing Mando's face.

How do the writers pull off this feat? By simply burning the helmet into Mando's narrative. The headgear is part of who he is. It's his creed, his culture. It's his backstory, and his future. His headgear becomes a kind of Schrödinger's space, naturally compelling the audience to wonder what's happening under the helmet. (Austrian physicist Erwin Schrödinger stated that if you place a cat and something that could kill the cat, such as a radioactive vial, in a box and sealed it, you wouldn't know if the cat was alive or dead. The condition of the cat was essentially unknowable until you opened the box, so until the box *was* opened, the cat was effectively both alive *and* dead.)

Picture a scene from The Mandalorian's "Chapter 15: The Believer." In the previous episode, Moff Gideon sent out four Dark Troopers who captured Grogu and took him back to Gideon's Imperial light cruiser, where Grogu was stunned and shackled. In "The Believer," Mando sends Gideon a threatening message. In a voice that is a masterful amalgam of cool and emotional, and vaguely reminiscent of Liam Neeson's iconic threatening phone call in the movie Taken, Mando vows to rescue Grogu: "You have something I want. You may think you have some idea of what you are in possession of, but you do not. Soon, he will be back with me. He means more to me than you will ever know." We don't need to see Mando's face. We can hear the strength of feeling in his words. We're so heavily invested in the story that we naturally place ourselves in Mando's position. This is made easier by Grogu's undeniable cuteness, so we easily imagine what Mando must be thinking and feeling, what his facial expression must be behind the dark visor.

This example from "The Believer" proves that we don't need to see Mando's face to know what emotions are playing out inside that helmet. We're invited to divine the drama in Mando's "Schrödinger's space" through other channels, including Pedro Pascal's voice acting and body language. We don't really need to see *any* character's face to tell us what to think in a scene. Consider a 1990 interview in the *LA Times* with British actor Michael Caine. Caine is quoted as saying, "Less is more. That's the hottest tip I can give any young actor. To do nothing at all can be very useful . . . Adopt a blank look if, say, you find your wife murdered, and the audience will project their own emotions onto your face."

We simply can't finish this section on the cinematic helmet without also considering the "Schrödinger's space" that is the headgear of Darth Vader, the original helmeted icon of *Star Wars*. Surely, Jon Favreau would have taken at least some inspiration from the example of Darth. The faceless Darth isn't quite the achievement that is Din Djarin, as Darth doesn't carry the weight of the story as Mando does. Nonetheless, Vader is a legendarily imposing figure, an iconic character in design, and excellently voice-acted, of course, by James Earl Jones, who brings the character to life. Vader also works so well that we don't need to see his face. In Vader's case, the characterization behind that helmet is helped by the infamous "Vader breath." That creepy, labored breathing is unnatural, somehow less than human, and speaks of some kind of alien disfigurement in Vader's Schrödinger's space. Maybe the helmet has never been the issue. The real problem has been the lack of creativity to successfully write the helmet into the story. Hats off once more to *The Mandalorian* in its creative embrace of the enigma behind that T-shaped visor.

The Mandalorian Helmet

How does Mando's helmet in *The Mandalorian* compare with its Greek counterpart? Given that the Greek hoplites had issues with visibility and hearing and, as far as we know, Mandalorian warriors do not fight in phalanx formation, how does Mando's helmet compensate for having that T-shaped visor? After all, wearing the helmets at all times was a sacred duty to warriors who followed the Mandalorian code, so they must have had fixes for living inside those helmets.

The first thing we should say about the Mandalorian helmet, even though it seems it's immediately recognizable throughout the galaxy, is that few of them were identical. Most appear to have had that telltale rounded crown, along with a T-shaped visor that totally covers the face. But, otherwise, there seems to have been a huge amount of customization and personal freedom in design, maybe surprisingly so given that the Mandalorians had a strict code and reverence for their armor. Perhaps in a culture where all members don their headgear at all times, customization was a convenient and flexible way of telling your friends apart.

It's easy to spot differences. Mando had a comparatively unadorned bucket of a helmet. Boba Fett painted his red and green. Sabine Wren, Mandalorian warrior and revolutionary leader during the early rebellion against the Galactic Empire, went one better and chose a kind of cosmic Power Rangers look, covering her helmet in a colorful orange and purple design. No doubt her flair for leadership was reflected in her radical helmet design, as even Sabine's no-nonsense mother, clan leader Ursa Wren, wore a strong canary-yellow Power Rangers look. Finally, at the other end of the bling scale, the very sober and humorless Armorer bedecked her helmet with nothing but punk-like spikes and metalwork, representing the material world of the Mandalorian: fist and boot, bullet and shell, blood and iron.

The fact that Mandalorians were feared opponents in battle suggests they had a whole array of tools and tech to supplement that helmet. Certainly, the tools and tech were enough to seduce renowned bounty hunter Boba Fett into adopting the Mandalorian armor, following in the footsteps of his father Jango Fett. Neither warrior otherwise subscribed to Mandalorian creed or culture, so when the Fetts managed to get their hands on a genuine Mandalorian helmet, what tech features seduced them?

One cool hallmark of the Mandalorian helmet is its telescopic zoom lens and range-finding feature. We see this function used by Mando several times in *The Mandalorian*. On many Mandalorian helmets, magnification looks to be linked to the targeting rangefinder, which appears as a sort of antenna when set to the stored position. We also see Mandalorians flip their rangefinder down to sit over their visor when they need to take a good look at their target. Not Mando, however. Mando's telescopic zoom function is controlled by a console he wears on his left arm. (Does Mando use Bluetooth?!)

Some Mandalorian helmets can even track a spaceship. Recall the first time we see Boba Fett in action in *The Empire Strikes Back*. In reply to Leia's "what now?" Han Solo has just divulged his plan to escape Darth Vader's pursuit by floating away with the *Star Destroyer* trash. As Han put it, "If they follow standard Imperial procedure, they'll dump their garbage before they go to light speed and then we just float away." But soon we see Boba Fett's ship trailing close behind. It's a pretty neat trick. Cut to the cockpit of *Slave I*. Boba Fett is seen, targeting rangefinder down, peering into some sort of scope. In all probability, he's piggybacking his targeting scope onto *Slave I*'s navigation system to suss out the *Millennium Falcon*'s course, beat Han to Bespin, and set up a trap in Lando Calrissian's dining room.

Then there's the helmet's thermal imaging function. We see plenty of evidence of this function in *The Mandalorian*. We learn that the helmet's thermal imaging capacity can penetrate reasonably thick walls and that the thermal setting is sophisticated enough to show fresh footprints while chasing after a quarry. In "Chapter 4: Sanctuary," Mando tracks seasoned warrior Cara Dune when he suspects that she's captured Grogu. Mando also mentions using his helmet to find footprints in "Chapter 8: Redemption," as he's scouring the tunnels for a way back to the Mandalorian covert.

Another function of the Mandalorian helmet is a feature that never appeared in the films. David West Reynolds's book *The Visual Dictionary of Star Wars, Episode II—Attack of the Clones* details Jango Fett's "pineal eye sensor," which sits at the crown of the helmet and allows Jango Fett to see behind him. What a useful feature for a bounty hunter, to be able to keep total tactical awareness during recon and combat (not to mention another way of dealing with the inadequacies of the Corinthian design).

In "Chapter 3: The Sin of The Mandalorian," Mando becomes an outlaw on the run. We see him use the powerful thermal imaging scope of his rifle to "look" through the walls of a building and spook a secret meeting between The Client and the ominously named Dr. Pershing. (Among other weapons, "Pershing" is also the name given to a solid-fueled, two-stage, medium-range ballistic weapon system named after US General John J. Pershing. As a result, one expects the good doctor to go "ballistic" at any moment; alas, it hasn't happened. Yet.) Moreover, Mando then pokes a button on the side of his helmet, which rather impressively enables him to hear The Client tell Pershing to "extract the necessary material and be done with [the Child.]" It's on the basis of this essential helmet function and its intel that Mando realizes the Child is in danger. He plans to rescue Grogu, even though doing so will sever his working relationship with the Bounty Hunters' Guild.

You would have thought the helmet's thermal imaging function would be enough for your average Mandalorian but, clearly, Mando is no ordinary Mandalorian. In "Chapter 8: Redemption," we see that, down in the sewers and suffering from "damage to his central processing unit" (IG-11's joke), Mando is able to attach a side-mounted flashlight to his headgear. We can only guess that, whereas thermal vision is useful for tracking warm bodies and peering through walls, it's not so useful for fixing your starship in the dark vacuum of deep space or fumbling your way around *Razor Crest* after a sudden power failure.

Introducing the Terrestrial Helmet

Do we have the tech to build our own Mandalorian helmet? Let's take a brief review of the most sophisticated helmets currently on the market. Given that Mandalorian helmets are fit for combat, our starting point is planet Earth's very own ballistic helmet. A ballistic helmet is the kind of headgear that protects the wearer's skull in combat, guarding against glass, shrapnel, bullets, and the like. They are often, less accurately, referred to as bulletproof helmets or tactical headgear.

Are ballistic helmets as robust as beskar? Not likely. But they commonly have several layers of ballistic materials to protect your cranium in combat or against other violent encounters. They're the kind of helmets we increasingly see police using in our troubled world. Many ordinary citizens who often use helmets, from building workers to bikers, could use a form of ballistic helmet in their line of work or play.

The best ballistic helmets are pretty tough. They are lightweight, coming in at just over one pound in weight, minus accessories. They have a high-quality chin strap as part of their retention system, something Mando seems to lack. They have a rail system for mounting additional gear such as night vision, cameras, or other optic devices. They are resistant to temperature, oil, and liquid, though not up to the standard of beskar. They often have an inner lining of multiple adjustable memory foam pads for added comfort and to reduce blunt force trauma from the like of bullets (which makes you wonder if Mando's helmet is padded). And they reduce back face deformation; this is the backside of the plate where the bullet is caught and prevented from penetrating your brain—hopefully! In short, when you're shot, the back face is deformed. All of which really makes you wonder how Mando survives the constant blunt force trauma of being a bounty hunter in battle.

The Terrestrial Helmet Build

How much do ballistic helmets cost? The basic build we have talked about so far will come in at around $600 at the time of writing. And this is minus accessories. Of course, the "best" combat helmet will depend on which branch of the military we are talking about, and the particular needs of the tactical situation at hand. But we are tying our colors to the mast of matching the Mandalorian helmet, so we shall go for the Enhanced Combat Helmet (ECH).

In order to approximate our best Mandalorian helmet, we need to add the following accessories to the ECH: 1, a telescopic zoom lens and range-finding feature (this could flip down over the ECH); 2, a device that piggybacks the targeting scope onto some kind of navigation system to track a spaceship (tricky); 3, a thermal imaging function (relatively simple); 4, a "pineal eye sensor" that sits at the crown of the helmet and allows the wearer to see behind him (also simple); 5, a side-mounted flashlight (utter child's play); and 6, some kind of device, controlled from the ECH, that enables the wearer to hear conversations through walls, as Mando did The Client's with Pershing.

Accessory 5, the side-mounted flashlight, is clearly already available and obvious, as ECH-type ballistic helmets already come with rail systems for mounting additional gear. And that means we can also tick off accessories 4, the pineal eye sensor, which is basically a camera, and 3, the thermal imaging function, as we can fit an infrared digital helmet night vision scope monocular to our rail system. We could try augmenting 3 with accessory 1, the telescopic zoom lens and range-finding feature. As for accessory 6, a device that hears through walls, there have been, for some time, espionage devices that enable such a thing. So that just leaves us with the relatively tricky accessory 2, the piggyback targeting scope.

Tracking Starships

How close can our terrestrial-built ballistic helmet come to pulling off Boba Fett's neat trick of tracking a starship? The short answer is, better than ever before. We live in a wired-up world. The speedy spread of networked devices means that listening and locating a quarry has become much easier, whether legally or illegally. Any new tech jacked into the web has the potential to become a surveillance device, whether or not its original function was benign.

Consider car tapping. For almost two decades, it's been possible to retrieve surveillance data from cars. Real-time audio and location data can be collected, as long as cops demand that vehicle tech providers hand the data over, and the tech providers comply. An example of this occurred in New York in 2014. The NYPD issued a warrant to track a target vehicle, a Toyota 4-Runner, ordering American broadcasting company SiriusXM, a satellite radio and telematics provider, to dish the dirt and hand over location data. The warrant instructed SiriusXM to "activate and monitor as a tracking device the SIRIUS XM Satellite Radio installed on the Target Vehicle for a period of 10 days." The Toyota 4-Runner was wrapped up in an illegal gambling enterprise.

SIRIUS XM complied. They switched on the stolen vehicle recovery function of its CVS (Connected Vehicle Services) tech, which is only available on a select number of its supplied cars. This means that the NYPD request was something similar to the police ordering Apple to hand over a particular phone's location data by switching on the Find My iPhone function. Other companies have also worked with the police in this way. General Motors (GM) has repeatedly complied with cops to provide not just location intel but also audio, where phone conversations are recorded when the in-car cellular connection was turned on.

Another way of tracking a terrestrial vehicle is using a GPS tracker. Such devices are fitted into vehicles to monitor their ongoing location. The GPS device receives data from satellites in the sky, repeatedly locating the moving vehicle and letting you know where on Earth you are via your chosen cellular network such as a web browser or phone app. And here's the thing. If you know where you are, a terrestrial Boba Fett can also find out, relatively easily, where you are. A GPS tracker is much more reliable than other tracking methods; unlike mobile apps, they don't need Internet access to work. It's all very well talking about car tapping, but how would our terrestrial-built ballistic helmet develop a GPS tracker for deep space? How on Earth would *that* work? It's time to talk about the quantum world.

Tracking Starships: A Quantum Detour

As if entering history with unerring precision, quantum theory was born in exactly 1900. In *this* Galaxy, that is, not in a galaxy far, far away. Quantum theory first occurred to German theoretical physicist Max Planck, who realized that he could explain the spectrum of energy radiated by an object if he ditched the traditional idea that energy is emitted continuously. He replaced it with the striking idea that energy comes in discrete units. Planck called these units quanta, after the Greco-Latin word for "how much" (as in quantity), and defined them in terms of the quantum of action.

A new enigma in terrestrial physics was born. Planck immediately knew that his quantum idea would revolutionize science, both on Earth and beyond. Not that he was, at the age of forty-two, particularly radical. Max was a doyen of nineteenth-century German high culture. But he hoped that his quantum principle would finally "show itself to be for the broadening and deepening of our whole knowledge in physics."

Planck's hopes proved prophetic. Forever in flux and evolving, quantum theory soon impinged upon almost every area of physics. And, as with all theories of science, quantum principles hold not just for this Galaxy, but for all galaxies. The quantum world is a very strange one. Georgiy Antonovich Gamov, Russian-born physicist and cosmologist, said that quantum theory was as if one could drink a pint of ale or no ale at all, but were otherwise stopped by a law of nature from drinking any quantity of ale *between* zero and one pint.

Things got even stranger than zombie Gungans. In 1927, the young German physicist Werner Heisenberg worked out his uncertainty principle. Werner found that one can measure either the precise position of a given particle, or its exact trajectory, but not both. Imagine watching an electron winging its way through a cloud chamber. We can work out the direction in which it is moving by recording its track through the cloud, but in the process of plowing through the chamber, the electron will have slowed by our actions, cheating us of the data about where it was at any given instant.

Even if we tried a different tack, nature would, so to speak, be ahead of us. Say we shine a light on an electron, kind of like taking a flash photograph of it. Our aim is to find out its exact position at a certain time. Yet, the light we use to take the picture will knock the electron off its seat, thereby robbing us of the exact intel of the location data we were looking for. Thus, we are limited in our knowledge of the quantum world. We can find only partial answers, whose details depend on the kind of questions we ask. I am trying hard not to sound too much like Yoda here, so please bear with me. We are still on the trail of that terrestrial-built ballistic helmet with starship tracking ability.

The thing is, way back in the history mists of atomic physics, wizardly scholars had thought that they could, theoretically, measure the exact locations and pathways of billions of particles and, from the ensuing data, predict where the particles might be at some time in the future. But Heisenberg proved this was false. As far as we can tell, we can never know all we need to know even about the behavior of a single particle, much less myriad particles, so we can never make predictions about the future that will be precise in every detail.

Characteristic Objects	Powers of 10 (meters)
Observable Universe (including quasars)	27
Superclusters of Galaxies	25
Clusters of Galaxies	24
Groups of Galaxies (including our Local Group)	23
Distance to Andromeda Galaxy	22
Milky Way Galaxy (diameter)	21
Distance to Orion Arm	19
Distance to Nearest Stars	17
Size of Our Solar System	13
Venus, Earth, and Mars	11
Earth-Moon Distance	9
Earth Diameter	7
Depth of Marianas Trench	4
Human Scale	0
Microorganisms	-4
Red Blood Cell	-5
DNA Structure	-8
Nucleus of a Carbon Atom	-14
Quarks	-16

Figure 4. Scale of the Universe: From Quark to Quasar

Then, there's the quantum leap. The closer our terrestrial wizards delved into the quantum world, the larger the uncertainty loomed. When a photon hits the atom of an element, it can send an electron into a higher orbit. But the electron seems to make this journey from lower to upper orbit instantaneously, without having traveled the intervening space. Way weirder than death-sticks. The electron orbital distances are also quantized. This means that the electron simply ceases to exist in one orbit, simultaneously appearing in another. *That's* quantum leap. It's a minuscule step, not a huge one, even though in common usage, "quantum leap" always seems to imply gargantuan advance. Unless we take quantum leap seriously, we simply can't explain the behavior of atoms.

If you're thinking *this quantum stuff is weirder than Ewoks on ice*, you're in good company. Danish physicist Niels Bohr, the other founding father of quantum theory along with Max Planck, once said, "If anybody says he can think about quantum problems without getting giddy, that only shows he has not understood the first thing about them." Niels is driving down to a deeper reality here. As human beings, we've grown up in a macroscopic world. We think in macroscopic terms. "Protons are like M&M's, light is like waves on the ocean, atoms are little solar systems," and so on. But these similes dissolve at the quantum level.

Just take a look at Figure 4 (page 114). Grasping the wee nature of the quantum world compared to the vastness of the observable Universe is a challenge most of us find confusing yet captivating (a little like Obi-Wan forgetting who R2-D2 is). Looking at Figure 4 should remind us that, with such differing scales, the quantum world behaves very differently to, say, the large-scale structure of the Universe.

Progress in the quantum revolution might have been painful early on. But we can now give thanks to early pioneers such as Planck, Heisenberg, and Bohr, for helping us escape the parochial nature of the way in which we live our daily lives, on Earth or on Mandalore. The wizardly scholars have delivered us from the affliction of a number of illusions. Chief among them is the notion that humans are somehow separated from nature, that the observations of science can be conducted with total objectivity. On the contrary, scientists are not the passive overseers they once believed themselves to be, shut off in an ivory tower of their own making, as remote as their view through the lenses of their deep-space telescopes.

The quantum world shows that on the infinitesimal level, and whether in Boston or on Bespin, our acts of observation are intrusive, disruptive, and influential. It's sometimes known as the "observer phenomenon," this idea that the very act of human observation can be the disturbance of the observed system. There's a striking story which helps illustrate the observer effect. Here it is (hint, I've changed the story just a little):

Mando is standing at the Crucifixion. Dumbstruck and open-mouthed, he can't help but stare at the scene. Perhaps the most famous in all of history. It was an expensive package. But his Time Travel Tour operator said it would be well worth the cash. Just a few points to remember: Do nothing to disrupt history. (Note to self: Don't tread on any butterflies.) And when the crowd is asked who should be saved, Mando should join in with the call, "Give us Barabbas!"

Suddenly, Mando realizes something about the crowd. Not a single soul from 33 AD is present. The mob condemning Jesus to the cross is made up, lock, stock, and smoking barrel, of tourists from the future.

The original story was called "Let's Go to Golgotha!" and it was a 1975 time-travel tale by science fiction writer Garry Kilworth. The crucifixion scene in the story isn't *passively* littered with people from the future. They've *actively* changed the outcome of history itself, by being present at the Crucifixion and observing it. The time tourists think they know the way history is meant to go. Rather than Jesus being set free, the crowd is meant to choose Barabbas, the bandit. But the decision only goes that way because travelers from the future (remember, Mando is actually from the past; "A long time ago in a galaxy far, far away . . .") are witness to the scene. Would Jesus have been set free instead, if they hadn't interfered?

We still live parochial lives. Even though macroscopic terms are inadequate to convey microscopic concepts, as wordsmiths we have little else in our arsenal other than words to describe the quantum world, so we paint pictures with our wits, and the medium we use is drawn from the world around us. Yet, the world as we see it is a lie, or, at best, an illusion.

On Mandalorian worlds, beskar was the toughest and most legendary "metal." On Earth, the chemical element osmium (Os) is the densest metal at all temperatures, denser by far than gold. Though osmium looks solid enough, it's composed almost entirely of empty space. Indeed, the nucleus of each osmium atom is so tiny that, if the atom were magnified a million billion times, until its outer limit was as big as London, its nucleus would be about the size of one of those black London taxi cabs. Consider another example, in which a pocket billiard ball collides with another. It's not so much that one ball strikes another, as much as the negatively charged fields of the respective balls repel one another. In fact, on the quantum scale, the pool balls are as spacious as Galaxies, and, except for their charged fields, they could, like Galaxies, pass right through each other intact.

Quantum Entanglement

All this counterintuitive weirdness of the subatomic world is preparation for quantum entanglement. The 2022 Nobel Prize in Physics was awarded to three scientists for their contributions to understanding quantum entanglement and advancing the field of quantum information. These three quantum physicists won the Prize for their experiments with entangled photons, in which particles of light become inextricably linked. The hope is that their experiments will act as the basis for an abundance of quantum tech, including quantum computers and communications.

Quantum entanglement is another bizarre, counterintuitive phenomenon, which explains the way in which two subatomic particles can be intimately linked to each other, even if separated by billions of light-years of deep space. Despite their huge separation, a change induced in one will affect the other.

The backstory of quantum entanglement began in 1964 when physicist John Stewart Bell suggested that such changes can be induced and occur instantaneously, even if the particles are whole realms apart. Bell's theorem is considered a crucial concept in modern physics, even though it contradicts other well-established principles. For instance, Albert Einstein had shown many years before Bell's theorem that information cannot travel faster than the speed of light. Fazed by the new physics, Einstein famously said that entanglement phenomenon was "spooky action at a distance."

Quantum Entangled Helmet Comms

The Wi-Fi in your home is powered by radio frequency (RF). And your Wi-Fi works because the RF waves carry data from a transmitter at one place in your home to a sensor at another. Your phone's GPS, and the communications on aircraft, work much the same way. The receiving sensors interpret the data in different ways. For example, a GPS sensor decides its location by noting the amount of time taken for a signal to arrive from a satellite. The more accurately the sensor reads this time delay, the more it can precisely work out its location.

In recent years, scientists have been working on ways to use quantum entanglement to provide a previously unheard-of level of precision with such communications. Terrestrial engineers and optical physics scholars are developing a combination of two techniques, radio frequency photonics sensing and quantum metrology, which uses entangled particles to make ultra-sensitive measurements. Their work entails transferring data from electrons to photons, then using quantum entanglement to increase the photons' sensing capabilities.

Conventional sensors transform data from RF signals to an electrical current, which is made up of moving electrons. But optical sensing, which uses photons, the quanta of light, to carry data, is far more efficient. Photons hold more information than electrons, so the signal has a larger bandwidth. And photonics-based sensing can transmit that signal much farther than electronics-based sensing, with less interference.

How does it work? Entangled particles are exactly that: entangled. So, as they are tied together, what happens to one particle will affect its entangled partner too. In the future, this could well be the way forward for tracking starships over cosmic distances and aping the antics of the legendary Boba Fett.

HOW WOULD AN ALIEN TECHNOLOGY FORGE BESKAR?

The Client: "The Beskar belongs back into the hands of a Mandalorian. It is good to restore the natural order of things after a period of such disarray, don't you agree?"
—"Chapter 1: The Client," Season 1 of *The Mandalorian* (2019)

The Client: "What exquisite craftsmanship. It is amazing how beautiful beskar can be when forged by its ancestral artisans."
—"Chapter 7: The Reckoning," Season 1 of
The Mandalorian (2019)

Supermaterials

In Mandalorian culture, beskar is a super material. Earth also has its share of super materials. We define our very history by the dominant materials of the day. The Stone Age gave way to the Bronze Age, which gave way to the Iron Age. The Stone Age lasted almost 3.5 million years, all that time making do with tools and weapons made of stone. It makes one realize what a revolutionary discovery metalworking was. Our own history's armorers, skilled workers who heat rocks to tease out metal, helped define entire epochs.

Much later, our ancestral artisans refined iron into steel. Again, changing the course of civilization. The last century or so has also been a story of super materials. Turning petroleum into plastic. Another revolution. Whenever we make materials that redefine the potential of the objects we can produce, we lead human history down a totally new track.

Super-material Beskar

Beskar looms large in the legend of the Mandalorians. In the very first chapter, Mando shares a brief exchange with his fellow Mandalorian, the Armorer, about culture and armor. While *The Mandalorian* itself doesn't share much detail on the cryptic importance of beskar as a super material, we know from stories in the *Star Wars* expanded Universe, including *Star Wars: The Clone Wars* and *Rebels*, that Mandalorian armor is famous in the Star Wars Universe.

Beskar armor can apparently withstand blaster shots. We witness this directly in *The Mandalorian*. It can even protect the beskar wearer from glancing blows from a lightsaber, which we witness in the "Legacy of Mandalore" episode of *Star Wars: Rebels*. Yet, despite its impressive robustness and resilience, beskar remains amazingly light. This allows the average Mandalorian to be agile and maneuverable. Moreover, the strength of beskar is such that some sets of Mandalorian armor last hundreds of years. In *Star Wars: Rebels*, Sabine Wren's armor is half a millennium old.

But is beskar a metal or an alloy? Reports seem to differ. Scientifically speaking, and if we are to work out how an alien technology might actually make beskar armor, the difference between the two categories is pretty crucial. Both canon and noncanon sources claim that beskar is also known as Mandalorian steel, based on Mandalorian iron, and thus must surely be an alloy, like terrestrial steel. Such sources go on to contradict themselves by then saying that "beskar was found only on Mandalorian worlds." If by this it's meant that only the Mandalorians were aware of some kind of secret process of making the beskar, fair enough. But if they meant that iron was found only on Mandalorian worlds, that's a contradiction. Here's why.

Forged by Fire: The Rise of the Terran Jedi

Consider Earth's history. Stone Age. Bronze Age. Iron Age. Fire played a key role in all three ages, but especially in the metal ages. The Bronze Age was a time of elite so-called palace civilizations, such as Ancient Egypt, when metal was the most important material. Our use of the word "metal" actually comes from the Greek and means "to search," which gives you a good idea of how scarce metal was at first. Indeed, that rarity explains why ancient humans first used metal only for ornaments and jewelry. But then came the techniques of fire, and metal smiths were able to alloy copper with tin to make bronze. Bronze is far harder than copper alone and makes far better tools and weapons. The innovation of fire and metal led to the rise of the first machines, such as the wheeled cart and the waterwheel, as well as city-states.

The Iron Age too was borne of fire. The very first iron to be fired probably came from meteorites or from iron ore as a by-product of gold-making. But when iron and its fire techniques became more widely known, it caused a democratic revolution that spelled the end of the great old palace civilizations and river empires of Egypt and Babylonia.

Iron making was easy, once you knew how. It needed only the simplest materials, wood and ironstone, and they were available almost everywhere, not "found only on Mandalorian worlds." Iron arose at the same time that people around the world became more mobile. The building of ships was a by-product of iron technology. Horsemen and sea peoples also had new iron weapons that challenged the power of the old empires, so power moved from the ancient elite civilizations to the more democratic republican cultures of Greece and Rome.

The change to an Iron Age economy inspired the rise of what we might call terrestrial Jedi. Instead of relying on ritual and superstition to fuel a worldview, a new paradigm emerged. The picture of the world that developed was both simple and material, based on everyday life and labor. The people who considered such matters were known as "sophists," or wise men. Only later did they become "philosophers," or lovers of wisdom.

These sages, like the Greek philosopher Pythagoras, often established religious orders that were also philosophic schools. The most successful leaders became political advisers to a democratic chief or tyrant (in those days, the word tyrant carried no ethical censure). They gave rational advice on every kind of topic. Indeed, it bestowed kudos on a regime to have a famous sage in tow. The irresistible rise of such Jedi-like philosophers was a global phenomenon. In ancient China, thinkers such as Confucius and Lao Tzu acted as political or scientific advisors. In early India, there lived at the same time the rishis and buddhas, Siddhārtha Gautama, the Buddha, being the most prominent. And in olden Palestine, the prophets and subsequent writers of the Wisdom literature, such as Ecclesiastes and the Book of Job, were alive.

Pythagoras: Terrestrial Jedi?

To give just one example of the link between the Terrestrial Iron Age and these Jedi-like philosophers, consider the example of the famous Pythagoras. It is useful to take some time to reflect upon the kind of ways that metal work may link with philosophy. Pythagoras and his Brotherhood had been among the first on planet Earth to realize the mathematical nature of the cosmos. They were among the first to determine that the Universe was ruled by number. This leads us to ask the question, why in ancient Greece does such rational thinking dawn? And what were the conditions of culture and economy that led to such a dawning?

The answer is actually linked to metal working. (Remember, science is like a recipe. It tells you how to do certain things, if you want to do them. As Yoda might say, science is not just a question of thought. It is the process of thought continually carried into action, constantly refined by practice. This makes science very potent. It is crucially linked to the means of providing for human needs.) According to legend, Pythagoras realized the link between music and math, and the cosmos and numbers, on hearing a blacksmith at work. In earshot of the sweet sound of the smith striking the anvil, Pythagoras understood that such harmony must bear some relation in mathematics. He spent some time with the blacksmith, examining utensils and exploring the simple ratios between tools and tones. The rest, as they say, is history.

At least two quirks of evolution made it much easier for terrestrial tech to progress and prosper. One was the prodigious availability of iron ore. The other was the subsequent accessibility of vast quantities of oil and coal to power the production of iron. Without iron and energy, we humans likely would not have gotten nearly as far as we have today.

Forging Terrestrial Steel

Now we see what nonsense it is to suggest, if beskar was Mandalorian steel, that the sourced iron for the steel would only be found on Mandalorian worlds. Consider our Solar System. Iron is found not just on planet Earth. Iron is also found on all of our rocky planet neighbors. Mercury's core, for example, has more iron than any other major planet in the Solar System. But it lacks iron in the mantle, while Mars is known as the Red Planet because it has twice as much iron oxide in its outer layers as our own planet. In fact, in the hunt for life beyond Earth, one of the key factors in determining whether a planet might be habitable or not is how iron-rich it is. Scholars now suggest that iron may influence the development of complex life forms. Iron is one of the most abundant elements in the Universe, along with lighter elements like hydrogen, oxygen, and carbon. And that means in the *Star Wars* galaxy, as in our own Galaxy, there should be abundant quantities of iron in its gaseous form.

Ignoring this contradiction about the iron for beskar only being found on Mandalorian worlds, how would their ancestral artisans, as The Client calls them, forge beskar from what we can only assume is iron metal? For starters, it's worth reminding ourselves that iron and terrestrial steel are two different things. Iron is cosmic, a pure element made inside stars. Stars are giant element furnaces, and when atoms are caused to collide in stellar interiors, new elements are forged—a process known as nuclear fusion. Such fusion is what ultimately created chemical elements like carbon or iron in the first place, those building blocks which make up life as we know it.

Meanwhile, steel is made on Earth. Steel is an alloy made from iron, which has most of the impurities removed. Steel is imbued with a consistent concentration of carbon of roughly 1 percent. It's the removal of impurities, such as sulfur, silica, carbon, and phosphorus, that makes steel much stronger than iron, as impurities weaken the steel. One way of making terrestrial steel is via an open-hearth furnace. Such steel is created from pig iron, limestone, and iron ore, which are placed into an open-hearth furnace and heated to about 1,600°F. The limestone and ore form a layer of slag that floats on the surface. The impurities are oxidized and rise up out of the iron and into the slag. Once the carbon content is correct, you have carbon steel.

An alternate way of making steel from pig iron is known as the Bessemer process. This entails oxidizing the impurities in the pig iron by simply blowing air through the molten iron in a Bessemer converter, named after its inventor Henry Bessemer, who took out a patent on the process way back in 1856. This must have taken a bit of nerve, as a system akin to this process had existed in East Asia since the eleventh century! Nonetheless, it's the heat of oxidation that raises the temperature of the process and keeps the iron molten. Finally, as the air passes through the molten pig iron, impurities unite with the oxygen to make oxides and form slag.

Most modern steel is made in what's known as a basic oxygen furnace. The process is speedy, about ten times faster than the open-hearth furnace. In basic oxygen furnaces, high-grade oxygen is blown through the molten iron, reducing carbon, silicon, phosphorus, and manganese levels. In addition, the use of so-called fluxes, chemical cleaning agents, helps limit the sulfur and phosphorus levels even further.

Here's arguably our first clue in the feasible forging of beskar. In the basic oxygen furnace, a variety of metals might now be alloyed with the steel at this point to render different properties. For instance, adding 10 to 30 percent chromium creates stainless steel, a product that is very resistant to rust. Whereas adding chromium *and* molybdenum creates chrome-moly steel, a product that is strong and light.

Forging Beskar

How would beskar be forged? According to numerous expanded Universe reference sources, beskar's provenance is iron ore mined on Mandalore and its Moon, Concordia. Apparently, carbon is next added to the iron in the foundry to enhance the steel's strength. Unfortunately, such carbon steel would be simply annihilated by a lightsaber strike. While the precise temperature of a fictional weapon like a lightsaber is merely a matter of academic debate, it's nonetheless clear that they are immensely powerful weapons that run astronomically hot. Some estimates put the heat of a lightsaber strike at roughly four times the surface temperature of our Sun. And if there's a metal that simply won't be able to handle even the briefest encounter with a weapon hotter than our nearest star, it's carbon steel, the lowest class of high-temperature metals.

Clearly, the ancient artisans of Mandalore used something other than carbon to forge their beskar. But even if they were to use metals such as tungsten or molybdenum, the resultant alloys could still only withstand below astronomically hot temperatures. Moreover, if these artisans forged a material that was incredibly resistant to heat, it would have been pretty tricky shaping it into anything, let alone armor and elaborate Mandalorian helmets.

Another challenge for ancient artisans of Mandalore is that the super materials able to withstand very high temperatures also tend to be rather brittle at lower temperatures. The last thing a Mandalorian warrior would want in combat is their beskar armor shattering on the mauling from a mudhorn, or fracturing when clenched between Rancor teeth. Besides, such alloys begin to oxidize at high heat, which means that a tungsten armor might cope with a lightsaber lunge, but it would certainly show a tarnished outcome from the encounter, of which we see no evidence, as Mando's suit seems just as shiny after a blaster hit as before.

This is what I suggest happened with those ancient artisans on Mandalore. Research in nuclear physics has shown that ultra-hot plasmas can be magnetically confined. And this means that beskar armor could be infused with some kind of magnetic barrier, a field that stops lightsabers from making contact on a nanoscopic level. In addition, the armor could have also been cooled by treating it with some sort of special coating on the surface, one that radiates heat energy away as light. Whatever the explanation, you can be sure that these ancient artisans changed Mandalorian history as much as ours have here on planet Earth.

WHY DOESN'T MANDO'S JETPACK USE REPULSORLIFT?

Bo-Katan Kryze: "A Mandalorian with a jetpack is a weapon."

—"Heroes of Mandalore," *Star Wars: Rebels* (2017)

Bren Derlin: "A lot of the younger soldiers thought we oughta be zipping around the battlefield on jetpacks. But one direct hit, and you're grounded. Maybe forever."

—*Star Wars: The New Essential Guide to Weapons and Technology* (2004)

Have Jetpack, Win Combat

In "Chapter 10: The Passenger," the Mandalorian jetpack is front and center of the action. We see Mando and Grogu whipping across the Tatooine sands on a speeder bike. But their return to Mos Eisley is interrupted in a ravine where three raiders, headed up by bounty hunter Kajain'sa'Nikto, manage to unseat our bikers by running a hidden rope across the ravine. Mando first uses his jetpack to steady his fall, then begins to fend off his ambushers. After managing to knock out two of his assailants, Mando is left to bargain with a third alien, who is holding a knife to Grogu's throat. Mando first offers to trade valuable cargo, such as Boba Fett's armor, but the alien demands the jetpack. Mando agrees, taking off the jetpack and carefully popping it down on the sand. The alien puts Grogu on the ground, grabs the jetpack, and runs off with it. With Grogu now safely cradled in his arms, Mando remotely activates the jetpack with his wrist gauntlet, sending the alien high into the air and dropping him down to his undignified death. The jetpack was Mando's wild card in turning the ambush around.

Meanwhile, in "Chapter 9: The Marshal," Mando and The Marshal, Cobb Vanth, use jetpacks to great effect in their battle with a krayt dragon on Tatooine. Mando and The Marshal fire up their jetpacks and fly to the top of a hill to confront the krayt dragon while Grogu watches from his cradle. Mando and Cobb fire at the dragon, forcing it underground. But the beast soon returns and rises up threateningly behind the general. On Mando's word, Cobb distracts the dragon by firing rockets at it. Mando now tells Cobb to look after Grogu before hitting Cobb's jetpack, causing The Marshal to fly away to the battle flanks. Mando then manages to force the dragon to swallow both himself and some activated detonators, and we finally sense a dramatic end of the battle as the krayt dragon reemerges from below the ground. Has Mando been consumed by the fierce creature? Not with a jetpack as a weapon. Mando flies out of its mouth with his jetpack, the dragon detonates into smithereens, and the townsfolk and Tuskens cheer.

"I'm Looking to the Sky to Save Me . . ."

In the Star Wars Universe, a jetpack is a personal aerial transportation gadget that enables its operator to fly into and through the air with great ease and mobility. Also known as thruster packs, jump packs, or rocket packs, many military forces use jetpack-equipped soldiers. But the jetpack is most closely associated with Mandalorian armor. The jetpacks have been a staple in *Star Wars* since Boba Fett's first appearance in the *Star Wars Holiday Special* in 1978. We don't see him using one in *The Empire Strikes Back*, despite wearing one, but we do see its live-action use in *Return of the Jedi*.

And, as Mandalorians are synonymous with jetpacks, we were never going to have to wait too long before we witnessed jetpacks on *The Mandalorian*. It was admittedly a few chapters in before Mando got his own jetpack, though by then we had already seen other Mandalorians taking to the air, and heard Mando respond by saying, "I gotta get one of those." Once established, we then get treated to the use of jetpacks in "The Passenger" and "The Marshal," see Bo-Katan with her jetpack in live-action form, and delight at the sight of Boba Fett retrieving his jetpack. Before we consider the jetpack, let's first take a short detour into the history of science fiction and the human fascination with flight.

Taking to the Air

In the history of science fiction, there are some pretty ingenious ways of taking to the air. Way back in 1657 (I kid you not!), Cyrano de Bergerac's book *Les États et Empires de la Lune* (*The States and Empires of the Moon*) was published. The protagonist in Cyrano's story takes to the air in the following way. Under the impression that the Sun "draws up" dewdrops, Cyrano suggests that one might fly by trapping dew in bottles, strapping the bottles around one's waist, and standing in sunlight. In such a way, the evaporating dew lifts Cyrano's hero to the Moon.

Another science fiction invention is the ramjet. The ramjet is a form of air-breathing jet engine that uses the forward motion of the engine to produce thrust. According to science fiction writer Arthur C. Clarke, it was Cyrano de Bergerac who invented the ramjet. Cyrano wrote:

> I foresaw very well, that the vacuity . . . would, to fill up the space, attract a great abundance of air, whereby my box would be carried up; and that proportionable as I mounted, the rushing wind that should force it through the hole, could not rise to the roof, but that furiously penetrating the machine, it must needs force it upon high.

Then there's antigravity. This idea of a force that opposes gravity became one of the great dreams of science fiction in the late 1800s. Typically, writers imagined all kinds of gadgets that enabled people or objects to hover, or be boosted about. There was even an antigravity principle known as "apergy" that was used to send spaceships to Mars in some early stories. (In one particular and far less romantic tale, an antigravity ointment is smeared on the hero's vehicle to enable space travel.) Buck Rogers even had an antigravity belt.

Antigravity is rife in *The Mandalorian* and *Star Wars*. Among many others, there's Luke's "hover" speeder on Tatooine, the sleek speeders on

Endor, and Jabba's *Khetanna*, the huge sail barge, with a crew of twenty-six and a capacity of five hundred passengers. Then there's Grogu's cradle, the so-called "hovering pram." The cradle is already iconic, that image of The Child's pram hovering as his guardian Mando stands by. In the Star Wars Universe, such cradles are used for shielding younglings, a valuable tool for guardians with children too young to protect themselves.

Flight: A Superpower

Let me tell you a story about the human yearning for flight. For some years now, I have toured parts of Europe with science shows such as *The Science of Stars Wars*, *The Science of Doctor Who*, and *The Science of Superheroes*. The shows would be presented as part of science festival programs in places such as Edinburgh, Belfast, Cheltenham, Hay-on-Wye, and the Royal Institution in London. In fact, it is at this last venue that my story about flight is based.

In the introduction to our shows, my colleague and I would try to ask thoughtful questions of the juvenile audience to act as a kind of icebreaker. For *The Science of Superheroes* show, that question was "if you could have a superpower, what superpower would you choose, and why?" This question would almost always elicit the same response in boys. The superpower they would invariably pick would be invisibility. And the reason they gave? Never good, always dubious. "Sneaking into the girls' changing rooms," for instance, or "raiding the local candy store without being seen," or (more ambitiously, given their age) "pulling off a bank heist."

Over the years of presenting the show, a variety of lesser superpowers were chosen. Superspeed and superstrength were popular with boys, obvious references to the likes of Flash and the Hulk. (Loki was always overlooked, as an obsession with immortality seldom presents in those so young.) The girls? There seemed to be a female fascination with flight. (There was one significant exception to this rule: a ten-year-old who declared her preferred superpower to be "purple power." When asked to elaborate, she simply explained that it was the "power to do purple." Only later did I realize she might have been alluding to Tyrian purple, the color associated with the ancient civilization of the Phoenicians, a color worn by Caesar and Cleopatra, and generally so connected with royalty and nobility that the word "porphyriogenatos" was coined to mean "born in the purple." Purple power, indeed!)

Repulsorlift Tech

The Mandalorian jetpack comes with a science fiction and antigravity pedigree. And this begs the question, why doesn't the jetpack use antigravity? In *The Science of Star Wars*, we point out that vehicles like Luke's *X-34* landspeeder certainly don't appear to use conventional tech like a hovercraft, as they clearly have no skirt of material to trap air underneath the craft. Nor do they use magnetic levitation (maglev), as maglev vehicles, such as terrestrial maglev trains, float upon magnetic fields that sit between train and track. And there's no track in the woods of Endor, and no rail in sight on the sands of Jakku.

The "secret," of course, is repulsorlift tech, the fictional technology in the Star Wars Universe that enables crafts to hover or fly above a planetary surface. Examples of repulsorlift tech include Han Solo's frozen carbonite form, floating through the corridors of Cloud City; the fact that, when Luke parks his landspeeder and switches off its engine, the vehicle continues to float, just like the carbonited Han; and every single craft lined up for the start of the pod race in *The Phantom Menace*.

The Jetpack

How is the jetpack meant to work? In *The Science of James Bond*, I report that the jetpack even crossed into mainstream film and fiction. The 1965 Bond movie, *Thunderball*, opens with Bond escaping with the use of a jetpack, which places Bond firmly in his own genre of "spy-fi." In *Thunderball*, the jetpack is the standard idea of a flying device, worn on one's back, that uses jets of fluid to propel its pilot through the air. Real jetpacks had been tried using a variety of tech but, in practice, there were problems. "Factual" jetpacks are far more limited than the fictional ones. Challenges such as gravity, the Earth's atmosphere, the low energy density of possible fuels, and the unsuitability of the human frame for flight make it all too tricky (no matter what mumbo-jumbo they use in *Star Wars* literature).

Take Google's word for it. Astro Teller, head of the Google X research lab, even went on record to say that they'd evaluated jetpacks and found them way too fuel-inefficient to be practical. They even cited alarming examples, with fuel consumption as much as 334 gallons per hundred miles! They were also considerably deafening for the pilot, being as loud as motorbikes to the wearer. So, since Google stopped developing them, wouldn't it make more sense for the jetpacks in *The Mandalorian* and *Star Wars* to use repulsorlift tech rather than tired and outdated ideas?

Repulsorlift Jetpack

One way of avoiding all this outdated tech is to make repulsorlift part of the jetpack build. The crucial ingredient, the factor that delivers the antigravity performance, is exotic matter. According to Einstein, all mass makes gravity by bending space. All that's needed is some kind of exotic matter that simply bends space the other way. Theoretically, exotic matter has negative energy, or negative mass (this simply means matter whose mass is of opposite sign to the mass of normal matter, for example, −10 pounds). In this way, for jetpacks as well as any other vehicle, exotic matter should create the opposite effect of gravity, and it could be used to cancel out the weight of the jetpack pilot.

The crucial bit of the jetpack build is how much exotic matter is needed to make Mando, for example, float. You simply measure the mass of your pilot, clothed and kitted out, and build into the jetpack an equal mass of the exotic stuff. The resultant mass of your jetpack is zero. With no effective resultant mass, Mando's jetpack won't be pulled down by the planet. And when he takes off his pack, just like Luke's speeder, Mando's jetpack will rest at whatever height he left it. With additional thrusters built in, Mando is now free to kill the task of being the best bounty hunter in the "parsec."

One last footnote. At the time of writing, no one actually knows what exotic matter is. On Earth, our definition is "any kind of non-baryonic matter." Normal matter is made of baryons, subatomic particles such as protons and neutrons. Exotic matter is simply made of different stuff. We're just not sure what that stuff might be. But why worry? They clearly know what exotic matter is in the Star Wars Universe, or else they would not have been able to build all those repulsorlift vehicles.

WHAT HUMAN HISTORY IS HIDDEN IN *THE MANDALORIAN?*

"The Stone Age did not end because humans ran out of stones.
It ended because it was time for a rethink about how we live."
—William McDonough, *Eco-Designs on Future Cities* (2005)

One-third to one-half of humanity are said to go to bed hungry
every night. In the Old Stone Age the fraction must have been
much smaller. This is the era of hunger unprecedented. Now, in
the time of the greatest technical power, is starvation an institution. Reverse another venerable formula: the amount of hunger
increases relatively and absolutely with the evolution of culture.
—Marshall Sahlins, *Stone Age Economics* (1972)

"I See Nothing but Death, and Chaos"

The Mandalorians had their own history, of course. A fearsome warrior culture of clan-based communities, they once fought the ancient
order of the Jedi during the fall of the Old Republic. And we know that
their warrior culture and distinctive beskar armor meant they were
feared throughout the galaxy, and once had political sway over two
thousand star systems. But empires inevitably fall, and so it was with
the Mandalorians.

Indeed, empire or republic; what's the difference? As Werner Herzog's
villainous Client character, a son of the Empire, says in season 1 of *The
Mandalorian*, "Judged by any metric—safety, prosperity, trade oppor-
tunity, peace—compare imperial rule to what is happening now. Look

outside: Is the world more peaceful since the revolution? I see nothing but death, and chaos."

The Empire was an oppressive regime, which squeezed the rights and freedom out of the galaxy's diverse citizenry. But years of war and attrition also left Mandalore inhospitable. Their world a relative ruin, the Mandalorians were forced to live within domed cities. And once the Mandalorian Civil War between the pacifists and the martial tradition-alists was over, and the New Mandalorians victorious, those who refused to give up their warrior ways were exiled to Concordia and were believed to have died out.

How Do Human Societies Evolve?

How do such human societies evolve? It's a question that has intrigued humans for a long while, but especially since the days of Darwin. His 1859 magnum opus, *The Origin of Species*, infused all contemporary progressive thought with a sense of history; scientific, social, political, and artistic. Science fiction became a popular way of voicing concerns for an increasingly urbanized humanity—the long road to Coruscant had begun! Sci-fi also helped us think about the social implications of evolution. And so this irresistible rise of the effects of evolutionary theory spawned around seventy futuristic fantasy novels in England alone between 1870 and 1900.

The past's idyllic vision of a static world became passé. In its place was mutability. Early sci-fi stories emphasized the ebb and flow of evolution, as the writers who told these tales found themselves living through unsettling changes to the fabric of society. After Darwin, the new paradigm was the process of becoming; the question of what would become of man. But also, where have we been? How has human society evolved in the past, and how does it all work? If we could understand the science of the laws governing societal evolution, maybe we could better marshal our futures.

Science Fiction Helps

The very best sci-fi has been a profitable means of talking about past human societies and how our changing notions of the past may be shaped and influenced by interaction with these fictional worlds. With characteristically black humor, brilliant sci-fi writer Kurt Vonnegut paid tribute to the role of science fiction in his wonderfully creative 1969 anti-war novel, *Slaughterhouse-Five*. The main character, recently witness to the firebombing of Dresden, considered that people were "trying to reinvent themselves and their Universe. Science fiction was a big help." Vonnegut himself used sci-fi to confront the growing horrors of the twentieth century. Perhaps if his readers could stand the unreality of science fiction, they could face a little bit more reality after reading *Slaughterhouse-Five* than they could before they read it.

By following the journey of Mando and Grogu as they seek to escape the assassins of the Bounty Hunters' Guild, seek out Jedi, and evade the butt-end of the imperial forces of Moff Gideon, what do we learn about human societies? Naturally, there are copious allusions to America's historical western frontier. But there are also hidden references to prehistoric, ancient, and early medieval societies in the material cultures and built environments of Mando and Grogu's tale.

When we look at history, whether human history or history in the *Star Wars* galaxy, it seems to be a mass of contradictions. Our focus is lost among the morass of imperial wars, invasions, revolutions, periods of progress, and days of decline. So many conflicts appear to swirl around in the chaos of social evolution, it feels at first as though it will be impossible to understand and explain events from a rational basis.

Many have tried to explain the laws that govern our evolving existence. Past theories have ranged from the idea of some kind of supernatural guidance of dubious deities to the Great Men theory, in which it was felt that people (men) in positions of power deserved to lead because of characteristics granted to them at birth that ultimately helped them become "heroes." Indeed, the supernatural guidance idea so dominated the outlook of "sophisticated" thinkers in Western Europe that when the "heretical" idea of evolutionary history first developed it was branded as being the work of the Devil's disciples.

Science of Perspectives

It is wise for the student of history to view things not as static entities but as processes in evolving life and motion. In other words, the world, whether it is Earth, Mandalore, or Coruscant, is not to be seen as a complex of ready-made things, but as a network of processes, in which all things go through a constant evolution of coming into being, passing away, or being transformed.

Take the Empire in *Star Wars*, for example. Since the get-go with the *Star Wars* franchise, the Empire has been a popular and enduring feature among fans. But the Empire wasn't a ready-made thing. It developed as an authoritarian and dictatorial evolution of the Galactic Republic. And it didn't last long. It came to power relatively quickly, lasted for a few decades, then eventually fell. The same happens to empires in real life, of course, though they last a little longer. The Roman Empire lasted just over 500 years, the Mughal Empire 331 years, and the Ming Empire 276 years. But all empires come into being and pass away, whether it is the British Empire, the Mongol, or the American.

Incidentally, there's an interesting take on the dialectic of empire in science fiction in *James Cameron's Story of Science Fiction*, a documentary series made by Cameron in 2018 for the AMC channel. In this series, Cameron interviews George Lucas about *Star Wars*, and the two embark on a fascinating exchange about the true nature of the Empire:

> **Cameron:** "You did something very interesting with *Star Wars*. If you think about it, the good guys are the rebels. They're using asymmetric warfare against a highly organized empire. I think we call those guys terrorists today. We call them Mujahedeen. We call them Al-Qaeda."
> **Lucas:** "When I did it, they were Vietcong."
> **Cameron:** "Exactly. So were you thinking of that at the time?"
> **Lucas:** "Yes."
> **Cameron:** "So it was a very anti-authoritarian, very kind of '60s against the man kind of thing, nested deep inside a fantasy."
> **Lucas:** "Or a colonial, you know, we're fighting the largest empire in the world."

Cameron: "Right."

Lucas: "And we're just a bunch of hayseeds in coonskin hats, who don't know nothin'. And it was the same thing with the Vietnamese. The irony of that one is, in both of those [wars], the little guys won. And the big, highly technical empire, the English Empire, or the American Empire, lost. That was the whole point."

Cameron: "But that's a classic—us not profiting from the lesson of history, because you look at the inception of this country and it's a very noble fight of the underdog against the massive empire. You look at the situation now, where America's so proud of being the biggest economy, the most powerful military force on the planet, it's become the empire. From the perspective of a lot of people around the world."

Lucas: "It was the empire during the Vietnam war. But we never learned, you know, from England or Rome or, you know, a dozen other empires, empires that went on for hundreds of years, or sometimes thousands of years, we never got it. We never said, well, wait, wait, wait, wait, wait, this isn't the right thing to do. And we're still struggling with it."

Cameron: "And they [empires] fall because of failure of leadership, or government often, and you have a great line, which is 'so this is how liberty dies.'"

Lucas: "We're in the middle of it right now."

Cameron: "Exactly. It was a condemnation of populism in a science fiction context."

Lucas: "That's a theme that runs all the way through *Star Wars*."

Cameron: "I think science fiction is so good at these kinds of social themes."

The "Stone Age" World of *The Mandalorian*

Let's go on a journey through an example episode of *The Mandalorian* to see if we can spot any evidence of the way in which human history has evolved. We spoke earlier in this book about "Chapter 14: The Tragedy," an episode in which Mando is on a mission to conjure Jedi. In said pursuit, he arrives with Grogu on planet Tython, the green, pleasant, and uninhabited rocky world situated in the galaxy's Deep Core. As we have already seen, our duo visits the ruins of an ancient Jedi temple that just happens to be an open-air hilltop megalithic monument, including a runic inscription upon the central spherical structure.

What kind of society and culture would have led to such architecture? When did they occur on Earth? Taking Stonehenge once more as our example, Stonehenge does not stand in isolation, but forms part of a remarkable ancient landscape of monuments from the later part of the Stone Age and early Bronze Age. In short, it was built over such a long period of time that society itself had changed before it was completed.

Our early human ancestors first appeared between five million and seven million years ago, most likely when a species of apelike creatures in Africa started to walk habitually on two legs. The posture of the body became more erect. The transition was also accompanied by a refinement in the flexibility and dexterity of the hand. Whereas other creatures with which we shared the Stone Age had different traits as means of survival, humans had none of these. To survive, humans had to develop the only resources they had, their hands and their brain. Through a kind of science (experimental trial and error), humans learned various skills that were handed down through the generations.

Communication became crucial. Imagine Stone Age humans hunting on a planet somewhere in the *Star Wars* galaxy, maybe ancient Mandalore. The ancient Mandalorians spy a group of beasts they'd like to fell and eat. But, today, a storm quickly brews up and scares the beasts away. The humans go hungry. But an important lesson is learned. They begin to recognize the signs in the atmosphere for a storm brewing: dark skies and a rising wind. So, on their next hunting trip, when the humans see the signs of a storm coming, they need to signal to one another that they need to speed up the hunt. They create a set of grunts and gestures to say just that: "storm is brewing, make haste." Such grunts were the start of speech, as humans learned how best to survive.

With communication came toolmaking. Mastery over nature began with the development of the hand, with tools and labor. This seminal moment in human history is also one of the most famous scenes in the history of the movies. In their 1968 masterpiece, *2001: A Space Odyssey,*

Arthur C. Clarke and Stanley Kubrick begin their odyssey with a small band of human-apes who are on that long, pathetic road to racial extinction. But after alien intervention, the human journey, that ultimately leads to the superhuman, starts when one of the human-apes triumphantly hurls an animal bone into the air, brandishing it as a tool and a weapon. In an amazing cinematic ellipsis, the bone immediately morphs into an orbiting satellite, and three million years of human evolution are written off in one frame of film. The message is clear. Primal bone technology marks the birth of modern humans. And humans and labor, through machine tech, are inseparable from the very outset. One can imagine a similar scenario in the *Star Wars* galaxy.

On Earth, humans were flaking crude stone tools roughly 2.5 million years ago, then an early human migration spread from Africa into Asia and Europe two million years ago. Humans have always been social animals. We were forced to band together and cooperate in order to survive. In stark contrast to most of the animal kingdom, humans developed the ability to generalize and think in an abstract way. Again, a similar pathway would have been forged among human civilizations in other Galaxies, as long as we ignore the huge unlikelihood of humans being there in the first place.

Consider the methods of early shamans, humans regarded as having influential access to the world of spirits. These methods were based on mimicking and sympathizing with the workings of the world. From archeological evidence of the cave art of Western Europe, it seems such shamans were already established in the Old Stone Age. Take the cave paintings of the Trois-Frères in the Ariège department of southwestern France. A painting there shows a shaman wearing stag's horns, an owl mask, wolf ears, the forelegs of a bear, and the tail of a horse. The value of such animagus behavior may have been to ensure a successful hunt. Primitive societies on other planets might have evolved similar techniques.

At the start, shamans would use likenesses, and later symbols, to perform an operation on something that would be considered transferable to the real world. An unbroken thread links these ancient symbols to those used with such success in modern science. Another feature of primitive thought that at some point separated itself from imitative or symbolic magic was the idea of the influence wrought upon the real world by spirits.

The idea of a spirit probably emerged from the reluctance to accept the fact of death. And death is common to all Galaxies.

The very first spirits would have been very worldly, indeed early spirits might even have been members of the tribe who had since passed on. But the idea evolved that it was necessary to win, or regain, the favor of a spirit, now god, by doing something that pleased them. The old idea of spirits split into two very different forms. First, it transformed into the idea of spirit as an all-powerful being, or god, that was to become the central figure in religion. Second, the notion of the spirit became divorced from human origin to become an invisible natural agent, such as the wind, or the assumed active force behind chemical and other crucial changes, or among Jedi, even the idea of the Force itself. This second idea of the spirit was to become hugely important in the evolution of the understanding of "spirits" and gases in science. Whereas all the other creatures of this Earth, and by extension other human worlds in the fictional *Star Wars* galaxy, simply use their environment to suit their needs, humans *make* the environment suit their needs. Humans master nature. Through our work and ingenuity, human labor changes the world in which we live.

The economic basis of the kind of society and culture that would have led to the type of architecture we see on Tython is very simple. Humans, in ancient times, would have been relatively rare creatures that roamed around in small bands, searching for food. In such nomadic lives, the task of food gathering reigned over all else. Terrestrial archeologists dubbed this period the Old Stone Age. And, always assuming we are happy to swallow the impossible pill that humans evolved elsewhere in the Universe, we can imagine that humans in other Galaxies would have also gone through a primitive phase of evolution. At such times, property didn't exist. All things made, produced, or collected were kept in common.

Human, Quit Those Roaming Ways?

If life for primitive humans in the *Star Wars* galaxy was anything like ancient human life on Earth, at some point humans like the Mandalorians may have stopped their wanderings. Rather than roaming around for food, they would have settled in a particular place, cultivating crops and domesticating animals. With more time on their hands, they would have been free to innovate new and better tools. A food-producing economy created, tribal communities would have sprung up around the globe. (In some parts of planet Earth, we still find tribes who prefer this way of life, this mode of existence.)

Of course, as we well know on our own world, the global economy would later move through periods of slave society, feudalism, and capitalism before it got to the kind of empire that Lucas tells tales about in *Star Wars*. So, here's an interesting question. Should there be other human societies on other worlds out in deep space, such as Mandalore, would the science of the laws governing societal evolution mean that those other societies went through the same stages we earthly humans have? Or would humans elsewhere remain "stuck" as hunter-gatherers or farmers? Given the way in which our own world has evolved, technologically speaking, you probably find it difficult to imagine a planet where humans are locked into hunting-gathering in perpetuity. However, as we have only the example of humans on Earth to go by, we don't really know if it is inevitable that human societies elsewhere develop through stages such as slave society, feudalism, and capitalism. Though, of course, the way that worlds are portrayed in *Star Wars* and in *The Mandalorian* certainly suggests strong parallels with our own world. In short, *Star Wars* suggests that such societal evolution is a universal law. In truth, we really don't know.

To take just one example of the progressions of human development, think about the agricultural revolution and the transformation from hunting-gathering to farming. With the benefit of hindsight, there's a very good reason *not* to turn to farming. In 2022, an international team of researchers analyzed DNA and took biometrics from skeletal remains of 167 ancient individuals found around Europe. The bones dated to around the time when farming emerged in Europe 12,000 years ago. The switch from hunter-gatherer lifestyle to farming took an average one and a half inches off human height, the study found. And shorter height is an indicator of poorer health, as it suggests ancient humans weren't getting enough nutrition to support adequate growth. So, these first European farmers probably experienced poorer nutrition and increased disease burdens that stunted their growth.

Yet, such communal societies also had their benefits. There was no private property, children belonged to the entire human tribe, and society was very democratic. There were no privileged elites, no police or coercive apparatus of the state, and human tribes were split into social units called clans, very large family groups that traced their lineage from the female line alone in so-called matriarchal societies. Again, if such societal evolution is a universal law, then maybe there were matriarchal societies in the *Star Wars* galaxy too.

Maybe this was the kind of society that built the architecture on Tython. A society whose natural simplicity seems idyllic compared to the complex world of Coruscant. A society where there were no imperial soldiers, no kings or regents, no princesses or prisons, no judges or lawsuits, and still affairs ran smoothly. Any disputes were solved by the entire community involved in them. The land belonged to the tribe. There were no poor or destitute, as the community knew their duties toward the aged, sick, and disabled. All were free and equal, women included. There was no place yet for slaves, nor for the subjection of foreign tribes. Calmer days before the ruin of Empire.

PART IV
CULTURE

WHAT'S THE GREATEST HEIST IN *STAR WARS* HISTORY?

The thing is, it is very undemocratic, because if you or I go to one of these banks along here somewhere with a few thousand dollars in a briefcase, if I'm a Brit and do it, I have to give a really thorough explanation. Bank manager may call in the police. I have to produce my passport. If I want to open an account, I have to produce a utilities bill and all of that. But, if Mr. Orloff comes to a bank here and says, "I am from Russia. I have millions and millions of dollars, please. And here is a letter from a reputable lawyer in Moscow. And here is evidence that I run hotels, casinos, whatnot," bank manager says, "What are you doing for lunch?"

—John Le Carré, in interview (2010)

Mandalorian Heist

Given that *The Mandalorian* walks and talks like a space Western, we shouldn't be surprised to see that Mando's story includes a heist. This heist, a steal of sorts, happens in "Chapter 6: The Prisoner." Mando teams up with a motley crew of malcontents typical of a Western. This crew don't carry six-shooters but do include a few cowboy cliches. There is a sharpshooter, though he turns out to be ex-Imperial marksman Migs Mayfeld. There's a grunting strongman in the Devaronian Burg. There's a getaway driver in the form of the droid pilot Zero (okay, I know Westerns don't have getaway drivers, but I'm doing my best here!). Finally, there's the ex-love interest. In Mando's case, it's the knife-wielding and mildly

145

hysterical Harley-Quinn–like Twi'lek woman Xi'an. It's Xi'an's brother Qin that our gang's hoping to jimmy out of a prison ship replete with security droids and a high-tech control room, a science fiction equivalent of those impossible-to-crack safes you see in some Western heists.

Star Wars is an imaginative destination. A mighty continent of tales to tell, so sweeping that it can't possibly be contained in just the dozen or so movies, hundreds of television hours and video games, and books sufficient to stock a small library. We've been mining its depths for almost half a century and have only just scratched the surface. While many tales tell of the mythic struggle between good and evil, the Star Wars Universe also embraces a wide array of other stories, a reflection of the world in which we live. Thus, tales have been told in the vein of sci-fi, sure, but also zombie horror and military thrillers. So, it's hardly surprising that Star Wars should also feature heists.

There are plenty of heists in Star Wars. Most famously, in Solo: A Star Wars Story, we witness the early adventures of Han Solo and Chewbacca as they join a heist within the galactic criminal underworld ten years prior to the events of the original Star Wars. Also, consider Solo's Eleven. Okay, so the story ended up with the title Star Wars: Scoundrels because Solo's Eleven, in the words of writer Timothy Zahn, was a "Star Wars version of Ocean's Eleven, [and] my original title Solo's Eleven [was] a little too close to the trademark!" Scoundrels is also cast in the space western mold. Consider the plot. Han Solo should really be basking in triumph. He'd just played his part as cocky space pirate and captain of the Millennium Falcon in a daring raid that demolished the Death Star and helped the Rebel Alliance land its first significant blow against the warring Empire. But Han lost the reward for his heroics and had little else to celebrate. Might this turn out to be a Pyrrhic victory? Particularly as Han is still deep in debt to the gangster gastropod, Jabba the Hutt. There's a bounty on Han's head, and if he can't cough up the galactic creds, he'll have to pay with his hide. All Han has to do is bust into an ultra-fortified bunker of a Black Sun crime syndicate underboss and crack the galaxy's most notoriously impenetrable safe (see what I mean about the Western trope of impossible-to-crack safes?). So, Scoundrels Assemble! Han calls up a motley crew of rogues who are both miracle workers and madmen, including his usual partners in crime, the loyal sidekick Chewbacca and the cunning Lando Calrissian. Their mission, should they choose to accept it, is to dodge, deceive, and defeat heavily armed thugs, killer droids, and Imperial lackeys alike, and pull off the heist of the century (as long as the century concerned is the one in which the year 0 ABY is located, or else known as year 3,653 after the Treaty of Coruscant).

In *Star Wars* comics, there appears to be an obsession with heist plots involving mystery objects. Multiple titles have featured heists for smugglers and dark Force users alike. In *Han Solo & Chewbacca #2*, our smuggling dynamic duo team up with Greedo for a heist with Jabba the Hutt, chasing down a mysterious object. Readers know the task goes awry, as Jabba ultimately pits Greedo against Han and Chewie for owing him credits. In *The Halcyon Legacy #4*, Lando and the pirate Hondo compete in a heist for a "priceless jewel." Finally, in *Crimson Reign #4*, the Knights of Ren are working for Qi'ra on a heist to break into a vault at Fortress Vader to snatch an unspecified prize, billed as being significant to the Sith.

Indeed, heists fit surprisingly well in science fiction. Little wonder when you consider that the heist has been a mainstay of cinema since the mid-twentieth century; they feature in books, television, video games, and across all media. After all, the heist narrative delivers. It gives readers and viewers many of the things they crave in a good story: daring action, outlaws versus "the system," a sense of panache, thrills, spills, and a blueprinted plan that comes together just perfectly. Our outlaws are the best at what they do; smarter, fitter, faster.

Heists are genre neutral. Sure, they mostly show up in modern action movies, but they also occur in fantasy and science fiction. Think of Christopher Nolan's movie *Inception*, a reverse heist story in which the characters must plant an idea in someone's head. Or mainstream heists that use science fiction tropes in their plots. For example, in *Ocean's Eleven*, a device is used that sets off an electromagnetic pulse to break into a casino vault. In *Mission Impossible*, all kinds of techno-gadgets are used to break into the CIA.

The Making of Obi-Wan

The best story of a heist in *Star Wars* began far less dramatically. In January 1976, sixty-one-year-old English actor Sir Alec Guinness was already a legend on and off the silver screen. He had fought against the Nazi Empire during the Second World War by serving in the Royal Naval Reserve and commanding a landing craft during the invasion of Sicily and Elba. After the war, he received his first Academy Award nomination for his role in the 1951 movie, *The Lavender Hill Mob*. He previously played Herbert Pocket in *Great Expectations* (1946), Fagin in *Oliver Twist* (1948), Colonel Nicholson in *The Bridge on the River Kwai* (1957), for which he won both the Academy Award for Best Actor and the BAFTA Award for Best Actor, Prince Faisal in *Lawrence of Arabia* (1962), and General Yevgraf Zhivago in *Doctor Zhivago* (1965). Huge characters, all. But the role of Obi-Wan was to launch Alec Guinness into another planetary stratosphere in all respects.

For, in that month of January 1976, Alec Guinness was offered $150,000 ($762,000 in today's money) to play Obi-Wan Kenobi. The character was to become legend. Jedi Master. Mentor to Luke Skywalker. Martyred at the hands of Darth Vader. Transcendentally guiding Luke through the Force in his fight against the Galactic Empire. Alec Guinness's performance as Obi-Wan in the original 1977 *Star Wars* film earned him a nomination for the Academy Award for Best Supporting Actor, still the only acting nomination for a *Star Wars* movie. The character of Obi-Wan has endured, revived by Scottish actor Ewan McGregor in subsequent movies and more recently in the miniseries *Obi-Wan Kenobi* on Disney+.

In a rare interview with the BBC's Michael Parkinson in 1977, Guinness was asked how he came to be involved in *Star Wars*. Alec Guinness's reply is so perfectly understated, and a little like the character of Obi-Wan himself:

> Well, it arrived as a script. I was just finishing a picture in Hollywood, another day to go, and a script arrived on my dressing table. And I heard that it had been delivered by George Lucas and I thought, *well, that's rather impressive because he's an up-and-coming*

and very respect-worthy young director. And then when I opened it and found out it was science fiction I went, "Oh, crumbs! This is not for me." And then I started reading it. It seemed to me the dialogue was pretty ropey. *But I had to go on turning the page.* And, I mean, that's an essential in any script—you have got to know what happens next, or what's gonna be said next. And I went on reading; I thought, no, I like this. If only we can get some of the dialogue altered. And then I met him; we got on very well and I found myself doing it. That's all.

Incidentally, another of Guinness's replies to Parkinson's questions is eerily relevant to the character of Obi-Wan. Parkinson asks Guinness why he warned American actor James Dean one week before his death not to get into his motor car. Guinness explained about a chance meeting he had with Dean in September of 1955: "My very first night in Hollywood, I met James Dean. It was a very, very odd occurrence . . . A woman I knew phoned up and said, let me take you out to dinner . . . and we finally went to a little Italian dive and that was full . . ." Then, Guinness says he heard feet running down the street toward them, and it was James Dean. Dean introduced himself and invited them back to the Italian dive from which they'd just been turned out. But, before they went in, James Dean said he wanted to show him his new car. Guinness explains what happened next:

The car was some little silver, very smart thing, all done up in cellophane, with a bunch of roses tied to its bonnet. And I said "how fast can you drive in this?" And he said "oh, I can do 150 in it." And I said "have you driven it?" and he said "no, I've never been in it at all." *And some strange thing came over me. Some almost different voice.* And I said, "Look, I won't join your table unless you want me to, but I must say something. Please do not get into that car. Because if you do," and I looked at my watch, I said, "if you get into that car at all, it's now Thursday," whatever the date was, "10 o'clock at night and, by 10 o'clock at night next Thursday, you'll be dead, if you get into that car." . . . And he was dead, the following Thursday afternoon. In that car. It was a very, very odd, spooky experience.

In another interview, Guinness reports that Lucas had pitched the Obi-Wan role as, in the words of Guinness, "a blend of the wizard Merlin and a samurai warrior. You can't beat that. And unlike most space fantasy, the characters George Lucas has created aren't cardboard. And the story is gripping. There's a quest, encounters with other forms of life, and conflict between good and evil." According to further reports, Guinness had privately dismissed the script to friends as "fairy tale rubbish," but his remarks to Michael Parkinson appear to cast doubts on that.

Obi-Wan Strikes Gold

In that 1977 BBC interview, Parkinson also said to Guinness of the first *Star Wars* movie, "And it's made more money than any other movie ever made, and you got yourself part of the action?" "So I'm told," Guinness coyly replied. But Parkinson was having none of this apparent coyness. He wanted to know to what extent Guinness had a part in the worldwide smash and grab of *Star Wars* profits.

The truth is that the guileful Guinness had pulled off a heist. Of sorts. He only accepted the part of Obi-Wan once Twentieth Century-Fox had not only agreed to his demand that they double their initial offer but also, and here's the inspired steal, *2 percent* of the back-end grosses. What are back-end grosses? Basically, it just means the gross. It's termed "back-end," as actors are usually paid up front, either as a lump sum or as wages. An actor may choose to wager on a blockbuster or push forward a personal project by forgoing some or all of the upfront pay in exchange for a share of the box office. A share of the gross, instead of net profit, can be a much better deal for the actor, as even with a sizable gross, there may be no profit. But Alec Guinness got both. And this was *Star Wars*, after all.

So, back to that 1977 Parkinson BBC interview with Alec Guinness. He asks whether Guinness got himself "part of the action." Parkinson gently pressed the great actor a little further, "two and a half percent, isn't it?!" Guinness's reply was a revelation. He says, laughing:

> No, no, no, no; not quite that. You want that story? I've tried to keep this dark, I don't know where this all sprang from. I think it was the [London] *Evening Standard* we blame for this. I had a contract. My agent said, I've asked for 2 percent of whatever, because we didn't think it would be making [much], you know. I've had a percentage on a film before and they lose money like mad if I have a percentage. And I said, oh fine, alright, 2 percent. And the day before the film opened in San Francisco, George Lucas phoned me and, again he's very diffident and very shy and quiet, and he has a funny little voice, and he said, "I think the movie is kind of going to be alright." I said, I'm glad, George. He said, "Yeah, the press quite like it . . . We're pleased with, you know, very grateful for the little alterations you suggested. And so we'd like to offer you another half percent, by making you two and a half . . ." But a matter of a few weeks later, in fact the day I saw the film (I'd just seen it the once), the producer, who again is a charming, delightful chap, I said, about this little extra something you were kindly offering; I wonder if I could have something in writing, just so that, you know, my agent and so on believes this. And he said, oh about the quarter percent, yes.

After the movie opened on May 25, 1977, Guinness recorded his stunned response to its reception in his diary two days later: "Splendid news of reaction to *Star Wars* continues to come in." And on June 3, "Am pinning my hopes on *Star Wars* percentage, which could bring me £100,000 or more if it does *Jaws* business, as predicted." Indeed, in the end, *Star Wars* far outstripped *Jaws*'s business. Lucas's film grossed over $300 million on its initial release.

So, here's the heist. From that percentage deal, Alec Guinness earned more than $7 million in an instantaneous lightsaber flash (estimated at $33 million today). And an estimated $100 million (that's one-tenth of a billion dollars!) by the time of his death in the year 2000 at the age of eighty-six. According to some reports, Guinness had also negotiated a small percentage of every *Star Wars* movie in perpetuity. So that, when new movies were released, Guinness earned extra fees for no work at all. Yet, despite all these fantastic fees for what amounted to a mere twenty minutes of screen time, Guinness was reluctant to return for a sequel.

In 2003, Alec Guinness's portrayal of Obi-Wan Kenobi was chosen as the 37th greatest hero in cinema history by the American Film Institute. And digitally reconstructed archival audio of Guinness's voice was used in the 2015 movie *Star Wars: The Force Awakens* and the 2019 movie *Star Wars: The Rise of Skywalker.*

Finally, for the connoisseurs among you, an author recommendation. We spoke at the start of this chapter of Alec Guinness's wonderful movie roles. Here's another of his performances to add to that list. Two years after playing Obi-Wan, Guinness was the lead role in the BBC's seven-part TV adaptation of John le Carré's novel, *Tinker Tailor Soldier Spy*. The suggestion at the time was that Guinness was able to play George Smiley, a master spy hunting down a Soviet mole in the British intelligence services, because Guinness was rolling in *Star Wars* money, allowing him some latitude in selecting his next roles. It paid off. Guinness's portrayal of Smiley truly is one of the greatest performances ever seen on the small screen. Check it out.

COWBOY BEBOP:
WHAT WESTERNS SHOULD ALL
MANDALORIAN FANS WATCH?

Han: "Hokey religions and ancient weapons are no match for a good blaster."
Leia: "It's not over yet."
Han: "It is for me, sister. Look, I ain't in this for your revolution, and I'm not in it for you, Princess. I expect to be well paid."
                                        —*Star Wars: Episode IV—A New Hope* (1977)

Mando: "I can bring you in warm, or I can bring you in cold."
                —"Chapter 1: The Mandalorian," *The Mandalorian* (2019)

Manco: "Alive or dead, it's your choice."
                                        —*For a Few Dollars More* (1965)

Space Western

The COVID-19 pandemic was one of the deadliest in history. As of March 2023, there have been more than 676 million cases of COVID and 6.88 million confirmed deaths. Around the globe, public health mitigation measures included travel restrictions, quarantines, and lockdowns. But in all honesty, my life as a writer changed very little. I was already, and remain, in permanent lockdown with my studies. However, one adjustment I did manage to make during those lockdowns was to embark on a huge schedule of watching as many cowboy movies as possible. I'm not exactly sure why I even did this. Maybe some weird connection was made in my mind between the wild days of the pandemic and the Old Wild West. I watched over one hundred movies, old classics like *Stagecoach*, *The Man Who Shot Liberty Valance*, *The Searchers*, *Red River*, and *High Noon*, as well as more modern Westerns like *The Assassination of Jesse James by the Coward Robert Ford*, *The Revenant*, and *Django Unchained*. Imagine my delight in first watching *The Mandalorian* and being able to place all those classic old cowboy films into yet another contemporary context.

Apart from their setting in deep space, it could be argued that *The Mandalorian* and the *Star Wars* franchise have as much in common with Western movies as they do with sci-fi. Just look at the tropes that feature in the franchise. Saloons. Bandits. "Gun" duels. Bounty hunters. Outlaws with a price on their heads. "Space" exploration as a "final frontier." And a wild hero who doesn't quite fit in, but can't quite leave, and who saves the day on the sandblasted borders of civilization, where the old ways and customs haven't yet died out. Frontier worlds where bandits of every type eke out a living in the shadow of a monolithic yet far-flung government. A liminal world where life is cheap, and the only law comes from the barrel of a blaster. The planets presented in *The Mandalorian* feel more like one-horse frontier towns than they do sovereign worlds.

The details go deeper. Consider the character of Han Solo. Han's initial cowboy costume, and his beguiling antihero gunslinger ways, also echo a Western influence. *Star Wars* movies began to provide the tropes and morals that Westerns no longer offered. Moreover, in the DVD commentary to *The Empire Strikes Back*, George Lucas describes the character of Boba Fett as an homage to the "Man with No Name." The Man with No Name was the antihero character played by Clint Eastwood in the Dollars Trilogy of Spaghetti Western movies, created by Italian director Sergio Leone. In *A Fistful of Dollars* (1964), *For a Few Dollars More* (1965), and *The Good, the Bad and the Ugly* (1966), Eastwood is instantly memorable for his poncho-wearing, gun-toting, stubble-growing image. Sporting a brown hat and tan boots, with a soft spot for cigarillos and a taciturn nature, the Man with No Name was chosen as the 33rd greatest movie character of all time by *Empire* magazine in 2008, and is the single most iconic cowboy in the history of cinema.

The Man with No Face

Enter *The Mandalorian*. Mando is the man with no face. And his tale is a look at life in a lawless part of the galaxy after the fall of Empire. He is an enigmatic stranger who comes to a small border town to kick ass. A man of few words, we soon see that we can learn a lot from Mando's actions. He is a clever combatant, schooled in an ancient warrior culture, and he wears the suit of a long-defeated army. He may hunt for coin, but he also holds to a code. We see the man of honor emerge when he breaks contract and turns fugitive from the Guild to protect the Child.

This is a classic Western trope known as the "wolf and cub" narrative. Essentially, you couple up a hardened badass with an innocent that needs a guardian in hostile territory. We also learn a lot about our warrior from the way in which he treats those weaker than himself. Grogu. The rookie bounty hunter. The Sorgan villagers. And Kuiil. Mando treats them all with respect if not kindness.

Thus, *The Mandalorian* is a series of classic Old West scenarios, set out like an odyssey, a gauntlet through which our protagonist must run. There's a heist chapter. A jailbreak chapter. A man vs. nature chapter. There's even a *Magnificent Seven* chapter. It's all about how Mando survives the tests. Through every chapter, more of his character is revealed. And, through his shining example, we learn about the Mandalorians and their way.

Mando is as close to Wild West as you can get without actually saying so. In the words of Pedro Pascal, "The Mandalorian is a mysterious, lone gunfighter in the outer reaches of the galaxy. Some might say he has questionable moral character, in line with some of our best Westerns . . . And he's a badass. He's got a *lot* of Clint Eastwood in him." Din Djarin's alias of Mando is teasingly close to Manco, the nickname given to Eastwood's character in *For a Few Dollars More*. Given *The Mandalorian*'s use of Western themes and tropes within a sci-fi framework, let's take a look at the best the genre has to offer, and select the "magnificent seven" best Westerns of all time.

The Dollars Trilogy (also known as the Man with No Name trilogy)

There is no better place to start than Sergio Leone's loosely connected trilogy of Italian Spaghetti Westerns. These Westerns gained the label "Spaghetti" due to their origins being in Rome and for having an Italian director. Beginning with *A Fistful of Dollars*, Leone launched an entirely new and highly successful movie subgenre, with *Fistful of Dollars* also launching, seemingly overnight, the film career of actor Clint Eastwood. *Fistful of Dollars* was followed by *For a Few Dollars More* and *The Good, the Bad and the Ugly*. The Dollars Trilogy involves quests for money and focuses chiefly on Clint Eastwood's bounty killer character. Though often called the Man with No Name, in each film he is in fact named Joe, Manco, and Blondie, respectively. The films not only established Sergio Leone as a director, but also refined the aesthetic style of the entire Western genre, inspiring two hundred other Spaghetti Westerns, half of which contain the word "dollar" in the title!

Leone's influence remains. His unique approach to genre moviemaking is imitated not only in *The Mandalorian*, but also in the likes of Quentin Tarantino's 2015 Western, *The Hateful Eight*. This abiding influence is down to Leone's masterful direction. The scope of his stories feels larger than life, yet relatable and gripping. The look of his Spaghetti Westerns owes much to its Spanish locations that acted as a backdrop to his vision of a violent and morally dubious version of the American Old West.

What's the best order in which to watch the Dollars Trilogy? While there was never any intention on Leone's part to make a Clint Eastwood Spaghetti Western trilogy, nonetheless the movies possess a rough time-line that works reasonably well for the moviegoer. Treat yourself first to the masterpiece that is *The Good, The Bad and The Ugly*, with its stark commentary on the brutality of the American Civil War. This acts as a prequel to *For a Few Dollars More*. Finish your viewing pleasure with *A Fistful of Dollars*. Incredibly, despite Leone's intentions, or lack of them, the trilogy doesn't feature any glaring continuity errors when viewed this way. You will see just where *The Mandalorian* was conceived.

Leone's films were a remix of homage and refinement. His movies paid tribute to conventional Westerns, but were also markedly different in plot, characterization, and mood. Credit is due to Leone for the entire look of modern Westerns. Before his great breakthrough, traditional Westerns boasted relatively plastic heroes and villains who looked like they'd more likely stepped off a catwalk than a cattle drive. The conventional Westerns had clearly drawn moral opposites, even down to the daft idea of our hero wearing a white hat and the villain wearing a black one. In contrast, Leone's characters were truer to life and morally complex, often depicting outlaws or "lone wolves," usually unshaven, sweaty, and dirty, with a strong whiff of a criminal backstory. (When you think about it, this revolution in genre presentation also fed its way into *Star Wars*. Older sci-fi movies present a pristine future, dull and prissy anal retentives in silver, flame-retardant jumpsuits. But the first piece of tech we see in *Star Wars*, Luke's landspeeder, is replete with rust. It coughs. It splutters. But it's an all the more believable piece of futuristic tech. This contradictory characteristic is one of the many things that set *Star Wars* apart from everything else. The Rebels always sported preworn tech.)

The characters in the Dollars Trilogy were morally ambiguous. Now apparently compassionate, now nakedly self-serving, they seemed to lurch between different views of the Old West, from an optimistic view of the frontier to a world of death and corruption. It is often said that Eastwood's antiauthority character and rebel style in Leone's Westerns resonated with the social changes of the liberal 1960s.

In *The Mandalorian*, Mando begins his journey as a nakedly self-serving bounty hunter; amoral, and seemingly unconstrained, the entrepreneurial bounty killers are, after all, obsessed with accumulating a personal fortune. Like Eastwood's character, Mando is also an individualistic gunslinger with little time for bureaucracy. But Mando transforms into a much more compassionate man when he comes across Grogu. In Leone's Westerns, relationships revolved rather cynically around power, and retributions were emotion-driven rather than conscience-driven, much like the behavior of the agents of Empire in *Star Wars* and *The Mandalorian*.

The influence of the Dollars Trilogy on *The Mandalorian* goes far beyond a single episode or visual cue. Just consider the storyline in *For A Few Dollars More*. This movie sees Manco reluctantly join forces with another bounty hunter, and it's a narrative *The Mandalorian* uses twice. In the show's first season, Mando was made to join forces with the bounty hunter droid IG-11, climaxing in the memorable season one finale when IG-11 sacrificed himself to save Mando and Grogu. In season two, Mando aligns himself with the return of the legendary Boba Fett to help repel Moff Gideon's stormtrooper attack as Grogu meditates on the seeing stone.

The music of Leone's trilogy was also iconic. The scores' composer Ennio Morricone was a master of Western gothic soundtracks. A legendary composer in his lifetime, Morricone eventually won an Academy Award for Tarantino's *Hateful Eight*, at the time becoming the oldest person ever to win a competitive Oscar. *The Good, the Bad and the Ugly* in particular is not only one of the best film scores in history, but also hands down the best ever Western movie score and has been inducted into the Grammy Hall of Fame. Think of the Old Wild West and one instantly hears that opening whistle on the wind followed by three guitar notes that insinuate danger is just around the corner. The theme is complex and features electric guitar, drums, bass ocarina, chimes, trumpets, whistling, and lyrics sung by a choir featuring weird words like "wah, wah, wah," "go, go, go, eh, go," "who, who, who." (Check out the live performance on YouTube of *The Good, the Bad and the Ugly* by the Danish National Symphony Orchestra. At the time of writing, it has 110 million views, and for good reason, not the least of which is the fact that there appears to be a man hanging from the rafters by a rope, over the heads of the orchestra!)

One instrument you don't expect to hear in a theme song for an action hero is the recorder, yet a soprano recorder has pride of place in *The Good, the Bad and the Ugly* theme. Before being an instrument of torture for parents of kindergarten kids, the recorder was considered a serious instrument in Renaissance and Baroque music.

When Jon Favreau hired Swedish composer Ludwig Göransson to score *The Mandalorian*, they talked about the type of movies that influenced the show. Particularly, Favreau talked to Göransson about Morricone and his scores for Leone's Spaghetti Westerns. Given that *The Good, the Bad and the Ugly* theme had become a pop culture leitmotif, what better way to invite your audience to think of classic Western tropes. All you need to do is play the opening bars from *The Good, the Bad and the Ugly*, and even people who've never seen the film will imagine the Old Wild West. How did Göransson prepare himself for the task set by Favreau? In Göransson's words, "One of the first things I did when I started this project was to order a kit of recorders."

Given that it is a recorder trill that triggers an audience to the almost Pavlovian response of "Western!" Göransson decided a similar sound would be superb for summoning the idea of "space Western" for *The Mandalorian*. Making sure he creatively recast the feel of Morricone's score, Göransson chose the lower sound of the bass recorder, as opposed to the soprano recorder used in *The Good, the Bad and the Ugly*. Göransson also slowed the trill a little, and the new iconic theme for *The Mandalorian* was born.

Once Upon a Time in the West

After directing *The Good, the Bad and the Ugly*, Sergio Leone decided to retire from Westerns but was lured back by an offer from Paramount Pictures. The resulting movie was the 1968 classic, *Once Upon a Time in the West*. Leone set out to make the ultimate Western. His mission was to celebrate the power of classic cinema, meditate on the making of modern America, and lament the decline of one of the most cherished film genres.

The movie opens with a scene of epic proportions. For the first fifteen minutes, there's little to no dialogue. The dust blows, and menace is in the air. A troika of gunslingers arrive at a train station in the middle of the Old West. They are hours from civilization, which is just as well. They are not civilized men, and killing is their business. All three gunslingers are ultimately insignificant to the bigger picture but help show us the ruthlessness of the westward land grab made manifest by the railroad's expansion west. (In *Star Wars*, the inevitability of the railroad and manifest destiny is replaced by the foreboding sense that the Empire is coming to each and every town, whether people like it or not.)

Once Upon a Time in the West shifts the focus from the Man with No Name to another dangerous gunslinger, Harmonica, played by Charles Bronson. In the narrative arc of Leone's critical engagement with the American Western, the age of the gunslinger is coming to an end, and

Harmonica ultimately leaves the boomtown, giving way to a nonviolent former prostitute Jill McBain, played by Claudia Cardinale, who now represents the American future.

The concept art that appears at the end of each episode of *The Mandalorian*, essentially the *Star Wars* version of the Man with No Name, draws clear inspirations from the Dollars trilogy, *Once Upon a Time In the West*, and, our next pick, Clint Eastwood's Oscar-winning film, *Unforgiven*.

Unforgiven

Clint Eastwood's 1992 movie *Unforgiven* is in the revisionist Western tradition. Sometimes called "anti-Westerns," revisionist Westerns subvert the myth and romance of conventional Westerns by using character development and realism to present a less prosaic view of life in the Old West. Eastwood's movie tells the tale of William Munny, a legendary but aging outlaw and killer who agrees to one more job years after he turned to farming.

The period of the archetypical Old West is generally believed by historians to have been between 1865, the end of the American Civil War, and 1890, the closing of the frontier by the Census Bureau. Like *Once Upon a Time in the West*, *Unforgiven* takes place in this period, when the Old West was becoming new. As in Leone's movie, the age of the gunslinger is drawing to a close in *Unforgiven*. Professional gunfighters are such an endangered species that journalists stalk them for scoops. Men like Munny, "a known thief and a murderer" who once slept under the stars, are now trying to support their families as farmers.

Clint Eastwood probably picked this period for *Unforgiven* because it echoed his own life. He began as a young gunslinger on TV and shot to world fame in Leone's Dollars trilogy. Eastwood was now a director himself, and Leone had died in 1989; he partly dedicated *Unforgiven* to him. The Western was moribund. Cinema audiences now preferred sci-fi and special effects. It was time for homage, or even elegy.

The opening shot is of the Sun setting, on Munny and the era he personifies. The movie's exteriors are widescreen creations that portray the vastness of the landscapes. But the daytime interiors are brightly backlit, with so much Sun pouring in that the figures inside are dark and ill-defined. A new civilized life indoors has made these people indistinct.

A rider named the Schofield Kid appears with an offer of cash for bounty hunting. At first, it's a firm refusal from Munny: "I ain't like that anymore, Kid. It was whiskey done it as much as anythin' else. I ain't had a drop in over ten years. My wife, she cured me of that, cured me of drink and wickedness." But Munny needs the money to support his kids, and ultimately, he turns once more into a fearsome man. The story plays out in traditional Western terms, with the corrupt sheriff, Little Bill Daggett, facing up against Munny, the righteous outlaw.

The long final act of the movie, a masterful sustained action sequence that shows Eastwood learned much from Leone, ends with Little Bill being fatally wounded. "I don't deserve this. To die like this. I was building a house," says Little Bill. Munny replies, "Deserve's got nothin' to do with it." But, of course, deserve has *a lot* to do with it. It's Munny who makes sure people get what they deserve. That unforgiving moral equilibrium, in which good ultimately silences evil, is at the heart of the Western, as at the heart of *The Mandalorian*.

The plot of *Unforgiven* is echoed particularly in *The Mandalorian*'s "Chapter 5: The Gunslinger." Mando comes across a young and hotheaded rookie bounty hunter by the name of Toro Calican. Calican is keen but rather feckless, recalling a running theme of *Unforgiven*: the incompetence of bounty hunters. In Eastwood's movie, the Schofield Kid is blind as a bat, and can't seem to hit a thing with his trademark revolver. And when William Munny makes to saddle up, he finds to his disgrace that he can hardly mount a horse anymore. Mando's relationship with Calican pays homage to a time-honored Western convention, that of the older gunslinger reluctantly teaming up with a young, hungry one. Mando and Calican's lukewarm liaison invokes the same tensions we witness between Munny and the Kid in *Unforgiven*.

The Magnificent Seven

Appraised as one of the greatest ever Westerns, the 1960 movie *The Magnificent Seven* boasted an ensemble cast that included Charles

Bronson (yes, him again), Yul Brynner, Steve McQueen, Robert Vaughn, and James Coburn, who are among the seven title gunslingers hired to protect a small Mexican village from a group of marauding bandits. This Old West–style movie is actually a remake of Akira Kurosawa's 1954 legendary Japanese film *Seven Samurai*, which was initially released in the US as *The Magnificent Seven*.

Our story really starts with Kurosawa. *Seven Samurai* is not just a great movie. It's also the source of a subgenre that has flowed through film and television ever since. And that's because, arguably, it's the first film in which a team is assembled to carry out a mission. *Seven Samurai* gave birth to its direct Hollywood remake *The Magnificent Seven*, but also to 1967's box office success, *The Dirty Dozen*, about a real-life WWII unit behind enemy lines, and Marvel's *Avengers*. *Seven Samurai* surely influenced the Primetime Emmy Award-winning episode of *The Mandalorian*, "Chapter 4: Sanctuary," and no doubt the countless war, heist, and caper movies released since 1960.

The samurai and the villagers who hire them are drawn from different castes and must never mix, according to society. Indeed, the villagers had earlier been hostile to the samurai, yet the bandits are a greater threat, so the samurai are hired; valued and resented in roughly equal measure. Why do the samurai accept this mission? For a daily handful of rice. Why do they risk their lives? Because "this is the way"; it's in the nature of what it is to be samurai.

In *The Mandalorian*'s "Chapter 4: Sanctuary," Mando seeks refuge on the port-in-a-storm planet of Sorgan; a small, forested, and swampy world in the Outer Rim. Actually, this planet is so small that it doesn't even have a spaceport, but it's a perfect place to lie low. Planet Sorgan is also where Mando meets fellow fugitive and former rebel Cara Dune. The two soon find that their services are needed. Those in need of their help run a small krill farm (similar to the rice paddies of the *Seven Samurai*) that has been ravaged by raiders. Thus, they wish to hire Mando and Cara to protect them. The pair set up a *Seven Samurai* type of plan: Train the krill farmers for the coming combat and try to defeat the raiders, along with their AT-ST (All Terrain Scout Transport) and its twin laser cannon.

Mando and Cara carrying out their plan of training the farmers and creating barricades around the village is straight out of *Seven Samurai*.

Director Bryce Dallas Howard gets all the geography spot-on and helps us grasp the plan. This may seem like an obvious point, yet forgetting to do that very thing was one of the major shortcomings of the recent big-screen remake of *Seven Samurai*, the 2016 remake of *The Magnificent Seven*. If you've seen *Seven Samurai*, you know how much of the editing and cinematography in "Sanctuary" feels directly drawn from Kurosawa's classic movie. Mando and Cara win the day, but the drama is too much, and Mando leaves with his young charge, taking to the stars in their bid for survival.

The English

Just when you think the last word about the Old West has already been uttered, up pops another masterpiece. *The English*, a revisionist Western TV miniseries, produced by the BBC and Amazon Prime, was released in November 2022. Written and directed by British filmmaker Hugo Blick, the film features an Englishwoman, Lady Cornelia Locke, played by Emily Blunt, who travels to the Old West in that pivotal year, 1890. Lady Locke is looking for revenge on the man she sees as responsible for the death of her son. There she meets Eli Whipp, played by Chaske Spencer, an ex-cavalry scout and native of the Pawnee Nation, to whom he is known as Wounded Wolf. Eli is on his way to Nebraska to claim the land he is due for his service in the US army, despite having been told that the white men will not honor their debt. The most unlikely pair, Lady Locke and Wounded Wolf discover a possible shared history.

The story that unfolds is six hours of the most gloriously haunting television. The Old West is portrayed as a lawless land where no one can hear you, or anyone else, scream. After a semimutual rescue and several bloody deaths, the fates of Wolf and Locke, along with his pilgrimage and her revenge story, become irretrievably bonded. Like Leone's Spaghetti Westerns, *The English* was filmed in Spain, which we witness in all its splendor as the pair cross the plains in search of their different ideas of peace. The love between Wolf and Locke, these two lost and troubled souls, becomes deeper and more tender in a way that eclipses mere romance.

Along the way, we meet a cornucopia of characters who wonderfully evoke the pitilessness of the Old West. In particular, the murderous Black-Eyed Mog, a previously pious, now villainous, Welsh woman who appears in traditional eighteenth-century Welsh costume and is engaged in a long-running feud with Native Americans in Kansas. Black-Eyed Mog illustrates Blick's point: how many of us would remain sane and morally sound in a lawless land where, for three score leagues at a time, no one could hear you scream?

Blick's rather simple story brings history into sharp relief. The dehumanization and wholesale slaughter of the Native Americans, upon whose hell the New World was born, is ever present in Wolf's tale. It's there in the charred remains of the massacres, there in the barbarism of old soldiers they meet, there in the tales of the people with whom they seek refuge.

This all goes to show that Westerns, whether space-based sci-fi like *The Mandalorian* or new revisionist works like *The English*, remain, to some extent, the backbone of American cinema. They may not represent 40 percent of all films made, as they once did, but love for these old films and reverent love for the form will undoubtedly continue to inspire creative people like Jon Favreau to reinvent the genre.

Mandalorian as Remix

The Mandalorian is a space Western remix. "Remix culture" is a term that describes a culture or subculture that endorses derivative art by combining or editing existing materials to make new creative works. The word remix originally referred to music, emerging in the late twentieth century during the heyday of hip hop. But remixing didn't begin with hip hop. Looking retrospectively, we can see that old British rock bands such as the Beatles and the Rolling Stones became hugely successful for "remixing" older American music, as Elvis Presley before them became world famous for remixing source material that was drawn from traditional Black blues musicians many years before. All these musicians were simply doing what artists do. Copying from others, transforming these ideas, and combining them with other ideas to create a new synthesis.

It's the same with *The Mandalorian*. Jon Favreau's brainchild is replete with remixes: boilerplate characters, salvaged plots and storylines, mooched icons, looted tropes, and just a few new ideas injected into the *Star Wars* canon. It's all part of Favreau's remix revolution. *The Mandalorian* is ultimately way more than the sum of its parts. It's a grand theft remix heist.

HOW DOES MANDO MAKE LIKE ROBIN HOOD?

Nemik: Freedom is a pure idea. It occurs spontaneously and without instruction. Random acts of insurrection are occurring constantly throughout the galaxy. There are whole armies, battalions that have no idea that they've already enlisted in the cause. Remember that the frontier of the Rebellion is everywhere. And even the smallest act of insurrection pushes our lines forward . . . The Imperial need for control is so desperate because it is so unnatural. Tyranny requires constant effort. It breaks, it leaks. Authority is brittle. Oppression is the mask of fear. Remember that. And know this, the day will come when all these skirmishes and battles, these moments of defiance will have flooded the banks of the Empire's authority and then there will be one too many. One single thing will break the siege.

—"Episode 12: Rix Road," Season 1 of *Andor* (2022)

Mando, the Masked Bandit

The saloons of the Old West. Watering troughs for trappers. Grogshops for gunmen. Gin mills for gamblers, cowboys, and outlaws. Some were little more than brothels, casinos, or opium dens. The first saloon, set up in 1822 at Brown's Hole, Wyoming, was to serve fur trappers. By 1880, roughly halfway through our Golden Age of the Old West, the huge growth of saloons was in full pelt. According to an August 1891 edition of the New York publication, *The Week*, in Leavenworth, Kansas, alone there were around "150 saloons and four wholesale liquor houses."

A dark figure appears at the door of such a saloon, silhouetted by the bright wilderness beyond. For a moment, the bedlam inside the saloon suddenly stops. The stranger's identity is hidden from the world. As slick with his wits as he is with his weapon, the Masked Bandit is far more than mere man. He is a folk tale, a myth, and an icon. And he can bring you in warm, or he can bring you in cold.

We have already compared *The Mandalorian* to Westerns such as *Unforgiven, Once Upon a Time in the West*, and *The Good, the Bad and the Ugly*. We have noted that *Star Wars*, as a franchise, can be read as a space Western. Yet *The Mandalorian*'s eponymous leading light, Din Djarin, isn't *just* a neat fit for cinema's cliché of a lone gunman. He is so much more than that.

Masked bandits like Mando can loom where you least expect them. It doesn't have to be a saloon. Atop a distant hill, perhaps. Among the Ripper Streets of old London or high above the dark city skyscrapers. Zorro, V (as in *V for Vendetta*), and Batman are all masked bandits, too. Indeed, the most famous masked marauders from the movies, though admittedly not all bandits, would also include the likes of highwayman Dick Turpin, gladiator Maximus Decimus Meridius, serial killer Hannibal Lecter, the Phantom of the Opera, Catwoman, Iron Man, Deadpool, and many other masked superheroes.

Masked Bandits: Traditional and Mandalorian

The masked bandit is a rather romantic icon from traditional cinema. But just as *The Mandalorian* subverts the Western, it also undermines notions of heroic masculinity in modern action movies. Conventionally, the masked bandit is a vigilante. They set themselves up in parallel to official authority. Think Bruce Wayne's Batman here, a masked alternative to the crime-solving power of the Gotham City Police Department. Or Spider-Man, who does much the same as Batman, but for New York City. Meanwhile, *The Mandalorian* spins a different story. Mando's anonymous heroism serves his creed, community, and "family." His is the way.

Typically, masked bandits are antiauthoritarian. They are outlaws, often members of a marginalized group. The masked bandit is not necessarily galvanized by vengeance or justice, but by a deep dedication to honor. Tale-telling and reputation are crucial elements, too. The legend must live on, their story must be told. Moreover, it helps both storytelling and reputation if the masked bandit is unquestionably cool. They've got to have some swagger, copious amounts of charisma upon which the enigma can be hung, and a mojo and allure beyond compare.

To some extent, *The Mandalorian* turns its back on macho heroics, those histrionics witnessed in modern vigilante movies like *The Avengers* or *John Wick*. Mando is a masked bandit apart. An antihero who loves a child. An outlaw who abides by a strict code of honor. A lone fugitive who actively fights against authority; not just because it's the right thing to do, but because his survival depends on it.

Mando and Robin Hood

Robin Hood is perhaps the most recognizable outlaw in the history of pop culture. The bandit who stole from the rich to help the poor. Though he didn't have a literal mask, the Hood used a multitude of disguises. A minstrel costume to steal Prince John's silver plates. An old man's garb to enter an archery contest and win the golden arrow. And perhaps his best disguise: the legend surrounding the outlaw's identity.

The legend goes way, way back in history. Known throughout the world, the story of Robin Hood remains maybe the best-known tale of English folklore. In pop culture, the name "Robin Hood" is frequently used for a heroic outlaw or rebel against tyranny. The first known reference in English verse to the Hood is in *The Vision of Piers Plowman*, composed in the 1370s by William Langland, shortly before Geoffrey Chaucer wrote *The Canterbury Tales*. The Sloane manuscripts in the British Museum have an account of *a* Robin's life, which says he was born around 1160 in Lockersley, probably modern-day Loxley, in South Yorkshire.

There have been myriad versions of Robin Hood in legend, fiction, and film over the centuries since. The stories appeal because he is a folk hero who stood up against, and often outwitted, people in power. The Hood has always been a man of the people, a North Country man, whether it be Yorkshire or Nottinghamshire, who takes up banditry to protect his community from the colonizing French who would seek to take advantage of them. Furthermore, his life in the forest, hunting and feasting with his fellow outlaws, and coming to the aid of those in need, reads like a great and noble adventure.

Who was Robin Hood? The origins of the man, the legend, and the historical context have been argued over for centuries. Research has found plenty of references to actual historical outlaws with similar sounding names that have been suggested as potential evidence of his existence. Indeed, as many as eight plausible provenances to the legend have been broached by scholars, including the idea that "Robin Hood," like "Mandalorian," was a stock alias used by, or in reference to, outlaws.

Like Robin Hood, Din Djarin is not the sole owner to whom the identity "Mandalorian," and associated "mask," belong. *The Mandalorian*'s lore details the way in which Mandalorians nurture continuity in their community, adopting and raising orphans loyal to the creed. The social structures that allow Robin Hood to be of the people also allow Mando to be of the people. Without a confirmed identity beyond the mask, value is ascribed not to the individual, but to the collective, created idea of who Robin Hood, or Mando, is.

Superheroes don't often do feelings. Their prospective relationships are seen as weaknesses, so we often see such heroes put their personal ties on pause for the good fight when they take up vigilantism. But masked bandits like Mando and Robin Hood need both adventure *and* community, thereby *The Mandalorian* subverts the idea that a hero can only devote their energies to justice *or* to family. *The Mandalorian* presents a protagonist who shows bravery while also nurturing the emotional bonds that are central to the story. The more rounded hero that Mando represents strengthens the conviction he shows while wearing the helmet and taking direct action. Mando, as a shining example of the masked bandit, displaces his ego so that his skills can be used for the good of the community he cares for.

Star Wars and *The Adventures of Robin Hood*

These similarities between Mando and Robin Hood should not surprise us. In *The Making of Star Wars*, legendary movie director Martin Scorsese spoke about George Lucas's work on the original *Star Wars*:

I remember George was writing *Star Wars* at the time. He had all these books with him, like Isaac Asimov's *Guide to the Bible*, and he was envisioning this fantasy epic. He did explain that he wanted to tap into the collective unconscious of fairy tales. And he screened certain movies, like . . . Michael Curtiz's *The Adventures of Robin Hood*.

One of the highest-grossing movies of 1938, Michael Curtiz's film is arguably the best early distillation of the Hood myths, the movie setting the tone for many future iterations of the character, codifying the myth into popular culture. Billed as little more than "the most glorious romance of all time!" the trailer declared:

England, in the galant days when history hung on the flight of an arrow, or the slash of a sword. When feudal barons ravaged the countryside to live in pomp and splendor. When one man alone dared challenge the might of his country's oppressors. Robin Hood, outlaw of Sherwood Forest, and his stalwart band, robbing the rich to feed the poor.

In that 1977 interview with the BBC's Michael Parkinson, Alec Guinness describes his experience of actually watching the first *Star Wars* movie in this way:

A marvelous healthy innocence. Great pace, wonderful to look at. Full of guts; nothing unpleasant. I mean people go bang bang and people fall over and are dead, but, you know, no horrors, no sleazy sex; in fact, actually no sex at all . . . a sort of wonderful freshness about it, kind of like a wonderful fresh air. When I came out of the cinema into Tottenham Court Road, I thought, oh Lord, London is awfully sort of gritty and dirty and full of rubbish!

Guinness could have easily been talking about *The Adventures of Robin Hood*. Curtiz's film, above all else, is simply fun. Swashbuckling Australian-born actor Errol Flynn beams ebullience as the beating heart of the movie. His self-assured strut could have easily inspired Han Solo in *A New Hope*.

Both *Star Wars* and *Robin Hood* were trailblazers in moviemaking technique, too. Pushing forward progress in new technology, George Lucas created Industrial Light and Magic to achieve his fantastic vision. And even though *The Adventures of Robin Hood* might not at first appear to be a technical marvel of moviemaking and boundary pushing, it was precisely that. In 1995, *Robin Hood* was declared "culturally, historically, or aesthetically significant" by the US Library of Congress and selected for preservation by the National Film Registry. Curtiz's film was an early example of the opulent history of Technicolor, made a full year before *The Wizard of Oz*, and using all eleven Technicolor cameras available at the time. *Robin Hood* also begins with an opening crawl, giving moviegoers an update of what has come before, and providing enough context for what they're about to witness. The *Star Wars* opening crawl is, of course, a signature feature of the title sequence of every numbered film of the *Star Wars* franchise.

The two movies are also linked through their music. *Robin Hood* was particularly noted for its Oscar-winning score by Austrian composer Erich Wolfgang Korngold. Korngold also wrote the iconic Twentieth Century-Fox fanfare that kept *Star Wars* company through its first thirty years. US composer John Williams recognizes Korngold, and his kind of music composition, as a specific influence on his career as a whole, and *Star Wars* especially. In a 1998 *Star Wars Insider* interview, Williams explains:

> I've been particularly fascinated with the émigrés from Europe in the 1930s—people like . . . Erich Korngold . . . they brought this tremendous European culture. In a certain sense, my colleagues and I are the artistic grandchildren of these men. We have been the beneficiaries of a rich tradition that grew up here in the early days of sound, in the 1930s and 40s.

Just as there are similarities between outlaws Mando and Robin Hood, so there are parallels between *Star Wars* and *Robin Hood*. As Alec Guinness might have said, *The Adventures of Robin Hood* is the sort of movie that moves you. It does your heart some good. Like *Stars Wars* and *The Mandalorian*, *Robin Hood* is a mythic tale of right and wrong, good and evil. It presents a moral standard in glorious Technicolor. It teaches liberal lessons and warms hearts. It shone a light so bright on the legend of Robin Hood, of medieval heroes and villains, that its evocation of the myth is still the gold standard, almost a hundred years later. In 2003, American film historian Roger Ebert wrote, "The ideal hero must do good, defeat evil, have a good time, and win the girl. *The Adventures of Robin Hood* is like a textbook on how to get that right." Little wonder it worked equally well for the proximal case of Han Solo and the more distal case of Mando.

HOW CAN WE TELL THE TALE OF *THE MANDALORIAN* IN SIX-WORD STORIES?

"For sale: baby shoes, never worn."

—Ernest Hemingway

"Streets full of water. Please advise."

—Robert Benchley

"Lovely spring weather, bubonic plague raging."

—Evelyn Waugh

"Bang postponed. Not big enough. Reboot."

—David Brin

Flash Fiction

Back in 2006, *Wired* magazine ran a most entertaining feature: the six-word sci-fi story. *Wired* took as their inspiration a famous six-word story, attributed to author Ernest Hemingway. As it happens, the claim of Hemingway's authorship stems from an ill-founded anecdote about a wager between Hemingway and other authors. As British science fiction writer Arthur C. Clarke recounted in 1991, "He's [Hemingway] supposed to have won a ten-dollar bet (no small sum in the '20s) from his fellow writers. They paid up without a word . . . Here it is. I still can't think of it without crying—*For sale: baby shoes, never worn.*"

Fictional works of extreme brevity are known as flash fiction, also called minimalist fiction. If cleverly cast, flash fiction can still convey character and plot development, even within such word economy. Some critics are of the opinion that flash fiction has a unique literary quality in its ability to suggest a larger story.

One example of minimalist fiction is known as Twitterature. Here, the skilled writer is expected to rise to the creative challenge of writing 140-character stories (upgraded to 280 characters in late 2017). In 2009, Penguin books published *Twitterature: The World's Greatest Books Retold Through Twitter*, which, it boasted, contained "over sixty of the greatest works of Western literature—from Beowulf to Brontë, from Kafka to Kerouac, and from Dostoevsky to Dickens—each distilled through the voice of Twitter to its purest, pithiest essence."

Twitterature includes the following gems:

- From Sophocles's *Oedipus the King*, "PARTY IN THEBES!!! Nobody cares I killed that old dude, plus this woman is all over me. Total MILF."
- William Shakespeare's *Macbeth*, "Home now. Lady Macbeth hot over coming power/my nads. she wants to kill Duncan TONIGHT. Can't tell if she's serious or just into dirty-talk," and later, "She was serious. Women, LOL."
- Bram Stoker's *Dracula,* "A damsel is bleeding from her ears and eyes! She's afraid of the Sun! Like a ginger! We must sort this out. She may be a vampire, but I can't tell the father. He wonders if her 'lady times' are just out of control."

Six-Word Stories

Another identified variety of flash fiction is the six-word story, and it was within this context and history that *Wired* magazine asked a number of sci-fi, fantasy, and horror writers from the domains of film, fiction, television, and games to have a go at penning their own six-word stories. Only Arthur C. Clarke refused to trim his tale ("God said, 'Cancel Program GENESIS.' The Universe ceased to exist."), but the remainder were even pithier masterpieces.

To get a taste for the very best stories suggested by these prestigious writers, take a look at the list below:

"Automobile warranty expires. So does engine."
                              —**Stan Lee (cocreator of iconic characters**
                                       Spider-Man, Iron Man, Hulk, etc.)

"Machine. Unexpectedly, I'd invented a time"
                      —**Alan Moore (creator of *V for Vendetta*, *Watchmen*,**
                                                    and many more)

"Longed for him. Got him. Shit."
                      —**Margaret Atwood (author of *The Handmaid's***
                                       ***Tale* and *Oryx and Crake*)**

"The baby's blood type? Human, mostly."
                      —**Orson Scott Card (author of *Ender's Game*)**

"TIME MACHINE REACHES FUTURE!!! . . . nobody there . . . "
                 —**Harry Harrison (author of *Make Room! Make Room!***
                                       which became *Soylent Green*)

"Tick tock tick tock tick tick."
                      —**Neal Stephenson (author of *Snow Crash* which**
                                       coined the term "metaverse")

"From torched skyscrapers, men grew wings."
                      —**Gregory Maguire (author of *Wicked*)**

"Epitaph: Foolish humans, never escaped Earth."
                 —**Vernor Vinge (author of *True Names*, the first story**
                                       to feature the idea of cyberspace)

"Lie detector eyeglasses perfected: Civilization collapses."
                      —**Richard Powers (author of *The Overstory*)**

Star Wars Six-Word Stories

I hope you now get the idea, but isn't it high time we also developed flash fiction for *Star Wars*? And why not some six-word stories based on *The Mandalorian*? To start things off, how about we try to encapsulate some of the major themes of the franchise in six words.

Perhaps we could summarize the Skywalker saga with "Family issues wreck far-off galaxy" or with "Lightsaber clashed, Vader fell, Skywalker redeemed." More controversially, how about "Leia: 'Baby's yours.' Luke: 'Bad news.'" The atrocities of the Death Star might be neatly encapsulated with "Autonomous non-satellite terrorizes distant galaxy," though that's a little too similar to our first story. As good versus evil is clearly a major theme in *Star Wars*, and Vader is one of cinema's greatest ever villains, we might also suggest "Vader's behind you! Hurry before he . . ." or more simply, "Darth Vader: Anakin Skywalker's downfall."

Some six-word stories might act as movie taglines—as long as they don't give too much away, of course. "Lightsaber ignited, battle begins, destiny awaits" is usable as a tagline, whereas "Jedi falls, Sith rises, balance lost" and "Skywalker's legacy lives on forever" go too far.

The Mandalorian Six-Word Stories

How do we tell the tale of *The Mandalorian* in six words? Well, that very much depends on our focus. Given one of the main themes of this book, a perfect start might be "*Star Wars* Western, lone gunfighter rides" or "Helmet on, heart hidden within." Historically, thinking about the tortured history of planet Mandalore, I might suggest "Mandalore falls. Details at noon" or "For sale: Mandalore. No intelligence detected." With respect to Mando's personal journey, I might offer "Din Djarin's journey, a lone wolf," "Foundling becomes warrior, protects Baby Yoda," "Boba Fett's legacy, Mandalorian's destiny," or "Bounty hunter turns protector, finds purpose."

Finally, how might we foresee some dramatic developments in future *Mandalorian* narratives? On the dark side, I might suggest "It cost too much, staying Mandalorian." More dramatically, and far less Disney, we might suggest "Vacuum collision. Orbits diverge. Farewell, Grogu" or even "For sale: baby cradle, recently exploded." But if Disney is true to form, it's far more likely to be something like "Unlikely duo saves galaxy from doom."

HOW DOES MANDO REVEAL A SCIENCE OF THE OUTLAW?

"When a man is denied the right to live the life he believes in, he has no choice but to become an outlaw."

—Nelson Mandela, *Mandela* (1994)

"The point about social bandits is that they are peasant outlaws whom the lord and state regard as criminals, but who remain within peasant society, and are considered by their people as heroes, as champions, avengers, fighters for justice, perhaps even leaders of liberation, and in any case as men to be admired, helped, and supported."

—Eric Hobsbawm, *Bandits* (1969)

Cowboys and Aliens

More than a decade before Jon Favreau conjured up the Western genius of *The Mandalorian*, he directed *Cowboys & Aliens*. The 2011 movie might have sounded daft but, hell, it had Han Solo in the cast. It even had James Bond! And with its extraterrestrial take on the Western, *Cowboys & Aliens* foreshadowed some of the aspects that Favreau was to revisit in *The Mandalorian*.

For example, Daniel Craig's cowboy character is taciturn and laconic; a strong, silent hero who can fight and kill better than he can talk. The invading aliens, armed with superior resources and firepower, are more than happy to crush anyone "weaker" than themselves in the pursuit of their agenda, much like the Empire. And Harrison Ford's character forms a Mando-like attachment to a young kid, teaching him debatably ill-suited things about survival.

What inspired Jon Favreau to make these movies? In other words, how did the lone cowboy hero become such a forceful figure in American culture?

Cowboys around the Globe

It's not as if the cowboy is peculiar to the United States. Communities of wild horse- and herds-men are still present in numerous regions all around the globe. Among these communities, a number are kindred spirits to cowboys. The gauchos on the plains of Latin America; the llaneros on the plains of Venezuela and Colombia; the vaqueiros of Brazil; and especially the vaqueros of Mexico, whose costume and vocabulary of the cowboy's craft make the basis of the modern cowboy myth: lasso, sombrero, bronco, mustang, lariat, and chaps (chaparro). There are analogous communities in Europe too. On the Hungarian plains we find the csikos; in the cattle-rearing regions of Andalusia we find horsemen whose flamboyant antics more than likely gave rise to the earliest meaning of flamenco; and on the plains of Ukraine and southern Russia we find the numerous Cossack communities.

There is no lack of potential for cowboy myths around the world. Indeed, the communities listed above come with their own created myths of macho and heroic semi-barbarian daring action, yet none have created a myth to rival that of the fortunes of the North American cowboy, a myth with momentous global reach and endurance. Why might that be?

The Origins of the Wild West

In European culture, including its North American diaspora, the "Western" and the myth of the cowboy is a recent mutation of an early and deep-set vision of the Wild West in general. This original vision of the Wild West could be argued to harbor two main factors: on the one hand, the dialectical collision between nature and civilization, and on the other hand, between freedom and social constraint.

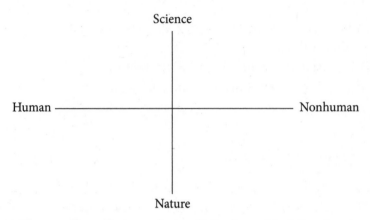

Figure 5. The paradigm of science fiction (© Professor Mark Rose)

Dialectical collisions are common to science fiction too. Science fiction is a way of exploring the relationship between the human and the nonhuman, and it helps us with our ongoing dialogue with science. Figure 5 (above) may help chart this ongoing dialogue. At times, as with the television series *The Last of Us*, science and the human are pitted against nature and the nonhuman. In such cases, the nonhuman comes in the form of a force of nature, which shatters the stability of the human world. In *The Last of Us*, an outbreak of a mutant fungus ravages the Earth, transforming its human hosts into aggressive creatures known as the "infected." In dystopias, such as the 1999 movie *The Matrix*, nature and the human are united in collision against science and the nonhuman machine world (the *Star Wars* franchise, of course, has a more nuanced attitude to bots and droids). As this classic *Matrix* movie suggests, sci-fi can sometimes characterize science as nonhuman and unnatural. The natural and organic human home world of Zion collides with the mechanical and scientific alternate world of the Matrix itself. According to this convention, utopias are imagined societies that are more fully human than the present. More often, however, science features on both sides of the human-nonhuman conflict. In *Cowboys & Aliens*, for instance, science is part of the nonhuman element symbolized by the invading aliens. They are agents of the void. They also embody science with their vast, cool, and unsympathetic intellects. Later, however, the alien invaders fall victim to a human plot that uses science and technical ingenuity to detonate and obliterate the alien mothership.

Lost Worlds

Until science fiction came along, the idea of the lost worlds of nature was practically unheard of. Before 1800, large parts of the planet were still unexplored and "uncivilized." As science and technology grew, and as modern civilization crept around the globe, travelers' tales became increasingly popular. A typical tale would first find our adventurer somewhere in the civilized world, usually London. Armed with a tall story, or an ancient scroll, our hero sets off to unknown or lost lands to find secret powers of great antiquity. Such are the classic lost world stories of the 1880s, such as H. Rider Haggard's *King Solomon's Mines* and *She*, the famous "She Who Must Be Obeyed." In the rapidly shrinking world of the twentieth century, no corner of the globe was left untouched. There were no more lost "arks" to raid. Writers found it ever more difficult to plonk their heroes in strange places that remained mysterious and unmapped. Heroes were shipped instead to other Earths. But with a dash of DNA chicanery, that didn't stop Steven Spielberg from mutating the quarrelsome dinosaurs from Edgar Rice Burroughs's 1918 *The Land That Time Forgot* into the menacing monsters of *Jurassic Park*.

The original Wild West epic is something of a previously unknown world. Plenty of white protagonists of such epics are misfits in or fugitives from "civilization," yet the main crux of their condition is that they are one of two types: visitors seeking something they can't find elsewhere, or explorers who form a symbiosis with nature.

The Power of the Cowboy

Given its young age as a nation, America can claim few art forms as its own. Jazz, certainly. And probably comic books, though that very much depends on what we actually mean by comics. But it's reasonably safe to say that, even though the cowboy has roots around the globe, we can add the Western to America's original art forms. America's particular take on the cowboy found its way into the world by two main avenues: the Western movie (a genre of film that was already firmly in place by 1909) and the hugely undervalued Western novel, which was as prominent in that past as crime fiction is to many of us today. The magic of the Western that so successfully carried the cowboy into the homes and hearts of people on five continents is still found in films like *One-Eyed Jacks* and *Red River, Stagecoach* and *High Noon*, and, of course, Sergio Leone's *Once Upon a Time in the West* and *The Good, the Bad and the Ugly*.

What was so cool about the cowboy? For one thing, they lived in a land whose frontier was globally visible and became a focus of the world of the nineteenth century. They were agents in the utopian drama of the living dream. In those halcyon days, frontier America seemed huge, wild, even unlimited in its potential. Moreover, the merely local American version of the Western myth was greatly maximized and globalized through the creative influence of American popular culture in the industrial world, along with the growing mass media that carried it.

Another influence was anarchy. This may be another reason why cowboys were considered cooler than gauchos or vaqueros and the rest. Burgeoning American capitalism had, at its core, a kind of built-in anarchism. Not just the laissez-faire nature of the markets, but the idea of the individual, free from the shackles of state authority. To some degree in nineteenth-century America, the West was regarded as a stateless society.

For example, think of the difference between the myths of the US and the Canadian West, and the contrast between the cowboys and the Mounties. The American story is a myth of a Hobbesian state of nature, an existence where each man lives for himself, only mildly diluted by occasional individual and collective self-help (Hobbes believed that when people have unlimited freedom, it leads to chaos and a war-like scenario.) The US myth is characterized by extreme competition, where no one looks out for another, where the land is populated by licensed or unlicensed gunmen, vigilantes, lone bounty hunters, and intermittent cavalry charges. In contrast, the Canadian myth is the tale of the imposition of state and public order as symbolized by the uniforms of a Canadian kind of horseman-hero, the Royal Canadian Mounted Police.

Given the nature of moviemaking for a mass audience, the prototype movie cowboy came in two flavors: the strong, silent, and retiring type, typified by the likes of John Wayne and Gary Cooper, and the cowboy entertainer type, typified by Buffalo Bill, and beautifully lampooned in the Coen brothers' 2018 Western anthology movie, *The Ballad of Buster Scruggs*. But the vehicles that propagated the cowboy myth died out. The Western novel was ousted by the crime novel, as private dicks shot the bounty hunters, and the Western movie was killed by TV.

Television Killed the Movie Star

By the 1960s, advertising had gotten its claws into the cowboy. Marlboro cigarettes showed the huge potential in gullible male identification with cowherders, who were being seen not as riding cattle but as gunslingers. The hackneyed use of the cowboy myth got so bad that in 1979, Ralph Lauren felt no shame in saying:

> The West. It's not just stagecoaches and sagebrush. It's an image of men who are real and proud. Of the freedom and independence we all would like to feel. Now Ralph Lauren has expressed all this in Chaps, his new men's cologne. Chaps is a cologne a man can put on as naturally as a worn leather jacket or a pair of jeans. Chaps. It's the West. The West you would like to feel inside yourself.

The tired and platitudinous cowboy clichés found their way into politics. Henry Kissinger stated in 1972 that "I've always acted alone like the cowboy . . . the cowboy entering the village or city alone on his horse . . . He acts, that's all." This political form would reach its apotheosis with President Ronald Reagan, the first president since Teddy Roosevelt to cultivate a Western identity, complete with Stetson and steed. Those Westerns that were still made became infected by Reaganism and became so peculiarly American and nonuniversal that the rest of the world lost interest.

Reinventing the Western

Take the example of Britain. Ever since those days of Marlboro country, the word "cowboy" has developed a different connotation. In Britain, a cowboy is a guy who appears as if out of nowhere offering a service, such as fixing your plumbing, but doesn't truly know what he's doing, or simply doesn't care other than wishing to rip you off; hence British culture harbors "cowboy plumbers," "cowboy mechanics," and indeed cowboys in most walks of life where a decent service would have been far more preferable. Whether the derivation of this fascinating usage is typical British irreverence in reaction to the cowboy stereotypes like John Wayne, or a political reaction to risible Reaganism, is perhaps rather academic. The point is the cowboy became associated with a fraud who would fleece you and fade off into the sunset.

Ironically, out of this European backlash against such tawdry images of the west came some of the greatest Westerns of all time. The Western may have been beautifully rejuvenated by Sergio Leone, but his was a worldview matured in the lore and the movies of the West, and skeptical of American cinema's invented tradition of the cowboy. Paradoxically, Leone's Spaghetti Westerns were at the same time very critical of the Western myth, whilst still showing how much demand remained among adults on both sides of the pond for old gunslingers and bounty hunters.

The Space Western

It's in the context of the history of the American cowboy myth, and its reinvention by Sergio Leone, that we should view the space Westerns of *Star Wars* and especially *The Mandalorian*. For the rich and powerful, American capitalism stood for the dominion of profit over state and law. Not merely because the state and law can be bought off, but because even when they can't be bought, they have no moral weight when compared to profit. This is exactly that selfish and corrupt spirit of the West so satirized by Sergio Leone in his Spaghetti Westerns.

In his movies, those with neither wealth nor power, such as the Man with No Name or Manco, possessed a spirit of anarchy that represented independence, the humble man's right to garner respect and show he's made of the right stuff. The archetypal and ideal cowboy hero, like Manco, was a loner, beholden to no one, nor someone to whom money was of the prime importance. Tom Mix, American movie star of many early Westerns between 1909 and 1935, said, "I ride into a place owning my own horse, saddle and bridle. It isn't my quarrel, but I get into trouble doing the right thing for somebody else. When it's all ironed out, I never get any money reward." Can't you so easily imagine Din Djarin saying the same thing? (But swapping in ship and planet for horse and place.)

The idea of the loner has great fecundity. Moviegoers find it easy to engage with the loner, to imagine themselves in his place. To be Gary Cooper in *High Noon*, or Sam Spade in *The Maltese Falcon*, or indeed Din Djarin in *The Mandalorian*, where you merely have to imagine yourself as one man, is far easier to grasp than to think of yourself as Moff Gideon or Darth Vader, as you have to imagine a whole bunch of people who honor and obey you. Being a dictator is just too much hassle.

The Mandalorian uses the power of the cowboy not only to harness the American past and the myth of an ultra-individualist society, the only society of the modern era without conventional roots, but also as a dramatic vehicle for dreaming about frontiers in space and the galaxy's unlimited opportunities.

A Science of the Outlaw

This book has been about outlaws. Bounty hunters and cowboys living outside the law, or else making their own law. In making *The Mandalorian*, Jon Favreau's creation rested not just on the more recent Western myths of the cowboy, but also on the centuries old tradition of the outlaw. Outlaws live large in our legends. Many such legends are relatively short-lived, but once in a while we get a Robin Hood. Robin Hood is the quintessence of outlaw legend. Like Din Djarin, Robin too may be a work of fiction, as no factual Robin Hood has ever been identified beyond dispute. This is unusual, as most other outlaw-heroes, however mythologized, can be traced back to some real person and place. If Robin Hood did exist, he must have prospered before the 1300s when his legend was first recorded in writing. His tale has thus been popularly told for over six centuries. Disney can but dream of a franchise that lasts that long.

The outlaw myth has surprising stamina. It has continued to thrive in highly urbanized societies such as ours, and especially in those countries that, at one time, possessed empty spaces and wildernesses to remind them either of an imaginary heroic past, a symbol of ancient and lost virtue, or freedoms in the face of the constraints of civilization and Empire. *The Mandalorian* taps into these universals of freedom, heroism, and the dream of justice, and on a far bigger stage, the Unknown Regions of a vast galaxy, rather than the parochial wilderness of a single country.

The legend of the outlaw Robin Hood is a tale of freedom, heroism, and social justice. Like Mando and Grogu in *The Mandalorian*, Robin's story too is about the fellowship of free and equal individuals, an invulnerability to authority, and the championship of the weak and cheated. In modern societies in which people live by subservience, as auxiliaries to machines of metal or moving parts of human machinery, outlaws like Robin and Mando live and die with straight backs.

Outlaws belong to a different kind of history. A *remembered* history, a folkloric history that stands in stark contrast to the official history of mere books. That's why outlaw legends like Robin and Mando still have the power to move people. As Czech writer Ivan Olbracht put it:

> Man has an insatiable longing for justice. In his soul he rebels against a social order which denies it to him, and whatever the world he lives in, he accuses either that social order or the entire material Universe of injustice. Man is filled with a strange, stubborn urge to remember, to think things out, and to change things; and in addition he carries within himself the wish to have what he cannot have—if only in the form of a fairy tale. That is perhaps the basis for the heroic sagas of all ages, all religions, all peoples, and all classes.

And that is why Mando is our hero, too.

INDEX